ഇൗ ൽ

CAREER
TRANSITIONS
for
CHEMISTS

ഇൗ ൽ

ᘓ ᘔ

CAREER
TRANSITIONS
for
CHEMISTS

ᘓ ᘔ

DOROTHY RODMANN
DONALD D. BLY
FRED OWENS
ANN-CLAIRE ANDERSON

AMERICAN CHEMICAL SOCIETY

Library of Congress Cataloging-in-Publication Data

Career transitions for chemists / by Dorothy Rodmann,
   Donald D. Bly, Frederick H. Owens, Ann-Claire
   Anderson

   p.   cm.

   Includes bibliographical references and index.

   ISBN 0–8412–3052–8 — ISBN 0–8412–3038–2
(pbk.)

   1. Chemistry—Vocational guidance. 2. Career
changes.

   I. Rodmann, Dorothy, 1930–   . II. Bly, Donald D.,
1936–

QD39.5.C24   1994
540'.23—dc20   94–38410
                                                         CIP

The paper used in this publication meets the minimum requirements of American National Standard for Information Sciences—Permanence of Paper for Printed Library Materials, ANSI Z39.48–1984.   ∞

Copyright © 1995

American Chemical Society

PRINTED IN THE UNITED STATES OF AMERICA

ABOUT THE AUTHORS

DOROTHY RODMANN brings to this publication comprehensive experience in the human resources field combined with private career counseling. Her knowledge of the organizational workplace coupled with several successful career transitions have given her insight into the strategies needed to pursue new career paths.

Dorothy's 25 years of work in the human resources field brought her to a senior management position at the American Chemical Society. She functioned as the director of Human Resources for the Society's Washington office from 1977 until late 1992. At that time she made the transition to the Membership Division to work more closely with member programs that matched her personal interests. Her present job title is Senior Advisor to the Director, Membership Division, and Consultant for Human Resources.

Dorothy's work in career services for the American Chemical Society began in 1987 at the New Orleans National Meeting, where she assisted in the design and presentation of the first career workshop for members. She has been involved in the design and presentation of other career workshops and continues to make presentations at local, regional, and national meetings.

DON BLY is a former manager in analytical science at Dupont, where he was responsible for research, measurement services, manpower, operating and capital budgets, and internal committees and symposia. At Dupont, Don also directed the corporate recognition and awards program and was involved with organizational effectiveness.

Currently, Don is self-employed as a consultant. He provides scientific expertise, educational instruction, critical examination of data, and audit of procedures for clients engaged in projects or other undertakings involving chemistry and its administration, especially within the broad areas of analytical chemistry, polymer chemistry, instruction, and employment. His educational background includes a B.A. degree from Kenyon College and a Ph.D. in analytical chemistry from Purdue University.

Don is a councilor for the Delaware Section of ACS and a member of its finance and other committees. He serves on the ACS Committee on Economic and Professional Affairs and is chair of the ACS Federal Policy Agenda Task Force. He has also been chair of both the Division of Analytical Chemistry and the Delaware Section.

FRED OWENS obtained his Ph.D. degree from the University of Illinois, working under Nelson Leonard on the chemistry of large-ring compounds.

In 1957 he went to work for Rohm and Haas Company in polymer research—his first transition. He rose to the position of section manager in plastics research, where he dealt with synthesis and applications of new polymeric systems; he obtained nine patents and authored or co-authored 10 other publications.

In 1976 came his second transition: He became manager of Research Information Services. During his tenure as manager of Information Services he regularly gave talks on careers in chemistry to middle and high school students, college students, and at student nights at ACS local sections.

He is a national ACS councilor and has served on a number of Council committees. He was a member of the ACS Joint Subcommittee on Employment Services for 17 years and was chairman twice. He is currently a member of the Employment Services Advisory Board. He is a career consultant and regularly participates at national and regional résumé reviews. He has given talks on résumé preparation at ACS section student nights.

ANN-CLAIRE ANDERSON is a staff associate in the American Chemical Society's Department of Career Services. She received a B.A. in English with a minor in biology from Trinity University in San Antonio, Texas, and an M.A. in English from Georgetown University in Washington, DC. Before coming to ACS, her primary focus was in adult education, including a position coordinating continuing education lectures and seminars for the Smithsonian Institution.

ଛୠ ଓଷ

CONTENTS

PART III: NONLABORATORY CAREERS

PART IV: OTHER CAREERS

APPENDIXES

໒ଓ ଓଓ

PREFACE

C urrent employment statistics indicate that a typical employee will hold about 10 jobs or positions with approximately five different employers over a lifetime. For chemists, this could be seven jobs or positions and three different employers. Because of the changes occurring in the workplace, today's chemist must be prepared to respond to an ever-changing array of opportunities and obstacles in his or her career. Chemists must be prepared at all times to implement one or more new career options. Job security in the new workplace is the ability to get another job.

This book is intended as a guide for chemists who may be facing a job change or career transition. Although the book stresses the importance of planning for a career change and performing self-assessments, it should be valuable to all chemists who are considering career moves. Even those in a crisis situation, who must find a job or make a career decision in the immediate future, will find the principles of transition useful because of the importance the process places on matching one's talents with the needs of an employer.

The career and job information in this publication was gathered by extensive interviews with members of the American Chemical Society (ACS) to clarify the knowledge and skills required for different chemical specialties. The personal assessment and transition information was assembled for and is used extensively in ACS career workshops and seminars, which are held at national and regional meetings. Although the information included is as comprehensive as possible, it is not intended to be exhaustive.

Parts II and III of the book outline several broad disciplines of chemistry as they are practiced. In researching this information, we drew from personal experience and telephone interviews with approximately 250 ACS members, who shared their employment anecdotes with us. Once we had compiled preliminary chapters treating each discipline, we forwarded them to experts in the appropriate fields for their feedback. The experts' comments assured us that we are presenting accurate and up-to-date material.

Each chapter includes a definition of that chemical discipline, information on what practitioners of that discipline do, lists of enchanced skills and knowledge they must possess, the places in which they typically work, and alternative chemical disciplines into which they may most easily move.

As the first book to provide current information for chemists on career transitions, this volume should be viewed as a combination resource manual and "how to" book. The chapters related to specific careers in the chemical sciences can be scanned and used as reference material. Chapter 2, entitled "The Transition Process," should be read in its entirety for each chemist to make the most of his or her own situation.

ACKNOWLEDGMENTS

We gratefully acknowledge the many people who contributed to the formation of this book.

Concepts for creating the book originated in the ACS Department of Career Services, directed by Mary L. Funke. Dr. Funke provided strong support for its formation, growth, and development, and we appreciate her personal contributions to the content as well.

Much of the information included in the book was obtained through experience and through interviews with experts. It is not possible to acknowledge by name the hundreds of contributors who were so helpful in providing us with information and keeping us on the right track. We thank them as a group for the many hours they contributed in discussing career transitions, personal assessment, and specialized fields of chemistry with us. We especially acknowledge those who contributed their biographical information, found in Appendix II.

Chapter 4 was prepared by Joan S. Burrelli, Senior Research Analyst in the ACS Department of Career Services.

The section on small business in Chapter 23 was prepared by Larry Bray, Birchard & Bray, Accountants; Ronald Gray, Chem Service Inc.; Lyle H. Phifer, Chem Service Inc.; and Jay Young, Consultant. The section in Chapter 23 on consulting was written by Donald J. Berets, consultant. We thank them all for their contributions.

Finally, we acknowledge the fine contributions made by our editor, Corinne Marasco, who helped bring continuity, readability, and style into the book.

Dorothy Rodmann, American Chemical Society
Don Bly, Consultant
Fred Owens, Consultant
Ann-Claire Anderson, American Chemical Society

ɞ ໒ଽ
PART I

THE CAREER TRANSITION PROCESS

॥ ॥

CHAPTER 1

MAKING IT HAPPEN

The responsibility for making a successful job or career transition rests primarily with you. *You* must determine what contributes to your job satisfaction and how you can best gain greater proficiency in the knowledge, skills, and abilities necessary for more effective work results.

Twenty years ago, organizations mapped out career paths for their employees, and for those of you who began your careers at that time, the thought of career self-management may be disconcerting. We believe that knowledge about the process of career management and application of its recommended strategies to an individual situation will help you make a better career choice. A good career choice means greater personal fulfillment from work and the likelihood of productive results for the employer. This is a win–win situation!

USING THIS BOOK

This book is intended to be your guide to doing your own career counseling. The sections on work functions, where your chemical knowledge is integral to the work process, are intended to help you discover connections between your background and work experience, and those skills that are required in the job market.

Interviews with a large number of chemists tell us that transitions may take any number of paths and may come about for a variety of reasons. Each transition is unique because of the different combination of skills, knowledge, and personal factors brought to the situation at that given

3052–8/95/0003$08.00/0 ©1995 American Chemical Society

time. In Appendix II, Transition Biographies, we provide examples of how some chemists made their own successful career transitions. Each of the biographies highlights different events that brought about career transitions and different personal factors that contributed to new career paths. You, too, will discover the personal uniqueness of the process as you work through your own transition. By highlighting events and personal factors that contributed to the transitions of some of your fellow chemists, we hope you will gain insight into some of the techniques that may be useful to you.

REASONS FOR CAREER TRANSITIONS

It may be surprising to you to learn that your movement from one job to another can be initiated either by you or your employer. The purposes for movement, however, vary. Sometimes a transition will occur because of luck, and you may view this as either good or bad depending upon your perspective. But transitions also occur for other reasons, some of which include those planned with a specific career objective in mind; those that are involuntary and caused by poor performance, poor work relationships, corporate downsizing, or strategic business change; voluntary retirement; outstanding employee performance (which causes an individual to be in demand); health considerations; and family needs. Regardless of the type of action precipitating the career change, the knowledge and use of the principles of making transitions can be helpful.

SEEKING CHANGE

The desire for change that motivates you to seek a voluntary transition is expanded upon in this section. Most of the reasons involve individual values, a topic that will be discussed in more detail later.

Reasons that may motivate a voluntary transition include

- You are looking for advancement or more autonomy, balance, challenge, or security in a job.
- You are seeking to broaden your skills and experience.
- You are seeking a new community or geographic location.
- You are seeking different benefits from those currently offered by your employer.

- You are dissatisfied with your current boss or management or job content.

RESISTING CHANGE

Whereas some chemists are busy preparing for transition and change, others resist change. We offer below a list of reasons that people resist change. You may want to examine this list to determine if any of them apply to you. If any of the points apply to your personal situation, an assessment of your current work environment and inhibiting personal factors can help you understand the circumstances causing the resistance.

Here are some reasons people resist change:

- They are comfortable where they are, and they don't want to put forth the effort that changing would require.
- They lack confidence to do something different.
- They lack the skills to undertake a change in direction.
- They lack vision; they can't see how things could be different.
- They deny the circumstances which suggest that change is needed.
- They believe that there is no legitimate reason for change.
- They fear the unknown circumstances of a new job.
- They are too burned-out to begin looking for another job.
- They have other personal issues preventing change.

WHEN ORGANIZATIONS INITIATE CHANGE

When organizations initiate change in the workplace, generally you can assume it is for business purposes and not directed at you. Some common reasons for organizational change are:

- Strong overall business growth or decline may require new deployment of staff.
- New staff levels are required because the organization is ignoring or swallowing up its competition.
- The organization is considering relocation.
- Changes in strategic direction force realignment of business units and major product lines.
- The organization is seeking mergers or divestitures.

- The organizational values are changing. (You may seem "out of step.")
- The business fails or becomes bankrupt.

If you sense that one or more of these things are happening within your organization, be alert to possible actions by the organization that could affect your career. These signs tell you that you should begin to plan accordingly. You can become part of the problem or part of the solution depending on the actions you take. This is a time for planning, self-assessment, and involvement.

Sometimes you may be surprised by unexpected actions or by your personal treatment by an organization. This may be especially true if a reduction in force, reprimand, or transfer is proposed. If you are alert to actions by your managers or the organization you work for, you usually will be given one or more signals that some impending action may be initiated in the near future.

When you pick up one of the signals, remember that this is a time for self-assessment. Begin to sort out your career options, and decide either to remain until more data are gathered or to initiate the first steps of your job campaign. Below are some of the warning signals to heed:

- Managers' communications to you become more negative, e.g., fault finding or nit-picking; they may appear less friendly; sometimes these managers are themselves being threatened and transfer such feelings down the line.
- Managers quit addressing your strengths, especially at appraisal time.
- Management is avoiding direct communication with you.
- Your requests for change in assignments are not pursued.
- Peers complain about you.
- You are excluded from important functions or meetings.
- Management gives the perception that you are not supportive of them or the organization.
- You are reassigned to another position without your understanding why.
- You are being left "out of the loop"; work is delegated elsewhere.
- Negative review results are a surprise to you, or your review is incomplete, not on schedule, or handled by other line management.
- You are passed over for promotion.
- You receive an unexpected direct reprimand.

MAKING A SUCCESSFUL TRANSITION

Having mentioned some of the more unpleasant things that can occur in the workplace, we want to balance that information by giving you some tips for making a successful transition should the need arise. You are in charge of the end result and in control of your future.

- Know yourself, your wants, values, and capabilities.
- Develop good communication skills—oral and written.
- Work on becoming more balanced in your life, both personally and professionally.
- Continually improve your interpersonal skills.
- Be a planner—try always to look ahead.
- Pursue education on a continuing basis.
- Maintain high performance levels in your work.
- Be motivated to act.
- Develop self-confidence.
- Be willing to risk change.

ജ‍ ‍ଓ
CHAPTER 2

THE TRANSITION
PROCESS

D escribing the transition process is analogous to describing the
mathematical   problem called the "Traveling Salesman Problem." In this problem, a salesman needs to visit a large number
of clients during an upcoming time period, and his challenge is to fore-
cast a course that covers the least number of miles or takes the least
amount of time. Alternatives are constrained by certain realities. For example, the salesman must be present at the clients' locations during
working hours, and he must allow time to talk with his clients.

The transition problem is similar to the salesman's problem. At any
time each of us is located in a specific job function—or unemployed
state—at a given location. Each of us also has a set of knowledge, skills,
and abilities in which we can demonstrate proficiency and a set of personal circumstances with which we have to cope. In accomplishing a
career transition, we may have to travel through several intermediate
steps and down several pathways (such as networking, obtaining additional education, or performing an interim job or apprenticeship) to
generate the needed knowledge, skills, and abilities required by the desired job. This journey is made all the more complex because unlike the
traveling salesman, all chemists in transition begin their journeys at a different point.

3052–8/95/0009$08.00/0 ©1995 American Chemical Society

THE SPHERE OF OPPORTUNITY

We have chosen to discuss transition by limiting the starting point to one position, which you will choose. To make a successful, long-term job transition, we will show you some possibilities; you will choose the specific starting point and connections that are right for you. Your guide is the model we call the "sphere of opportunity."

Chemists possess a set of core attributes that are independent of their areas of specialization (Figure 1). Those attributes that characterize us as being different from one another, as chemists, are contained in the particular body of knowledge we possess, in the different day-to-day tasks we perform, the degree of proficiency we demonstrate in the core skills, and in those specialized skills required by our particular chemical specialty. Chemical specialties are discussed in the next sections of this book.

To make the transition from one chemical specialty to another, you must possess the core attributes. You need to take a personal inventory to determine which of these core attributes are strong and which are weak, and to determine which need improvement or development for the new job. To transfer from a bench career into a nonbench, science-aligned career, the core skills also play a significant role, but their emphasis changes. Some of the core skills become dominant in the day-to-day activities required for proficiency in the new specialty, and some play a less important role.

Skills/Capabilities	Knowledge	Personal Characteristics
Analytical reasoning	Math	Curious
Creativity	Meaning of numbers	Goal-oriented
Computer capability	Validity of measure-	Self-starting
Data analysis	ments	Good at playing on
Experimental design	Record-keeping	teams
Problem-solving and	Basic laboratory	Honest
decision-making	knowledge	Dedicated
Reading and compre-	Specific chemistry	Flexible
hending chemistry		Good at communica-
		tion skills—oral and
		written

Figure 1. Core skills of chemists.

Figure 2 represents the sphere of opportunity. The myriad jobs held by or open to chemists covers the surface. One way you can envision this is to imagine the jobs as locations on the surface of a sphere. From each job location on the surface there are paths running to the center, like spokes on a wheel. The paths represent the core skills and attributes of all chemists. These skills and attributes are carried down from one job to the center and out again to the next job. The core skills are basically the same for all jobs. However, each job may emphasize a different mix or subset of the core skills and attributes.

At the center of the sphere is another zone, the location for personal assessment and judgment. This zone contains all of your personal factors, your values, your preferred communication style, and those additional factors that may influence movement, such as salary level and need to relocate.

For the purposes of illustration, positions A, B, C, and D in Figure 2 represent different jobs on the surface, and J, which stands for the

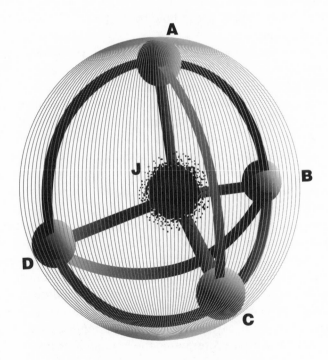

Figure 2. The sphere of opportunity.

judgment or assessment area, is the zone at the center. The absolute locations of the jobs on the surface and their relative locations one to another have no meaning. What is important is that they are separated by space and connected to one another two ways: (1) by curved lines on the surface and (2) by paths from the surface to the center of the sphere. One can travel from any position on the surface to another by two paths, by going down through J and out, and by traveling across the surface of the dome. To make a successful transition from one job to another, such as from A to C, you must use both paths.

To make the journey from one job on the surface of the sphere to another at a different location on the surface, it is always necessary to carry your core skills and attributes with you and to travel through the center, which is your assessment and judgment area. An assessment of the knowledge and skills required by the desired position and of any gaps existing in your current portfolio of skills, knowledge, and personal factors that could hinder your ability to make a transition is essential. The levels of fit that you determine will guide you, whether to travel the new path or to remain either in your current job or in the assessment area for some time, while you consider how to deal with obstacles in reaching your new position.

The surface link between any pair of jobs on the sphere connects the specific technical knowledge and skills required for each position. Unless this link is made, you cannot make a transition successfully. To connect one job to another successfully, you must determine what specific body of knowledge and skills each job requires and determine what you need to do to make the connection. Often this task requires obtaining additional education or experience. Traveling down and out through the center is only half the process of getting there; it is necessary, but not sufficient. You must also develop the additional knowledge and skills required by the new job.

In summary, making a job transition requires (1) that you obtain and develop a set of core attributes that you carry with you through life; (2) that you assess these core attributes along with your personal factors each time you make a transition; and (3) that you connect the special knowledge and job skills required by your existing and desired positions. These processes can be envisioned with the help of the sphere, the paths representing the core attributes, the center representing the personal factors, and the surface representing the job requirements.

THE ZERO-SUM LINE OF OPPORTUNITY

The sphere of opportunity affords a means to visualize the complexity of the transition process, and shows that job transitions are often complex. Simplistic approaches often fail to consider important elements. The zero-sum line of opportunity, illustrated in Figure 3, is a less complicated approach to transition, but do not ignore the personal assessment features, which are so important to the transition process.

Transition requires making choices and decisions. In achieving a career transition, you may have to travel through several intermediate steps or down several pathways (such as developing a network, obtaining additional education, performing an interim job or apprenticeship) to generate the needed knowledge, skills, and abilities required by the desired job. With the sphere of opportunity, you must transfer your core skills down and out the center path from one job to another, making note of

---

1. Write down the credentials (formal and informal knowledge), skills, abilities, values, communication style, and any other important attributes that you possess.

2. Write down the credentials, skills, abilities, values, communication style, and any other important attributes required by your desired job.

3. Subtract the two, i.e., note the differences and write down the credentials, skills, abilities, values, communication styles, and other attributes that you need to alter or develop.

4. If changes are required, determine how, or if, you will accomplish them and in what time frame.[a]

5. Make the changes and bring your credentials, skills, abilities, values, and communication style in line with those required by the desired job.

6. This result is the zero-sum difference between all your attributes and those required by the new job.

---

[a]Refer to the discussion on decision-making. Keep in mind that needed skills can be gained by training and education; but values differences are more difficult to overcome. Learn how to recognize and use communication styles as described in the section on communication.

Figure 3. The zero-sum line of opportunity.

what is being emphasized and what must be strengthened or abandoned. You must also assess your personal factors and you must obtain new knowledge and skills. And, by the surface path, you must compare the specific knowledge, skills, abilities, and tasks that are performed in each of two areas, your current job and your potential new job. By doing these things, you will determine what is needed for transition. All of these activities are also required with the zero-sum line of opportunity. We hope these illustrations help you to envision the process.

80 03

CHAPTER 3

PERSONAL ASSESSMENT

W hy is personal assessment esential to a successful job transition? Credentials or knowledge, both formal and informal,
provide a basis for demonstrating your competence to an employer. Your skills are the tools you use to implement or demonstrate the
knowledge you possess. Your values determine your fit and staying
power in a corporate culture, and communications help you to establish
rapport and influence the actions of others at all levels in an organization.

In attempting to accomplish a transition, you must penetrate the new
work environment with competence in each of these personal assessment areas. What this requirement means, and how to do this is discussed in the following sections.

SKILLS

Every individual possesses a unique collection of skills and capabilities
that is useful on the job or in obtaining a job. This personal collection
comes from each person's education, training, apprenticeships, hobbies,
and other activities.

Throughout this book we refer to a set of core attributes for chemists
(See Figure 1 in Chapter 2). These attributes form a unique collection,
not for all individuals, but for all chemists. They are the collection of
skills and other characteristics expected, consciously or unconsciously, by
employers of chemists. To be successful on the job or in changing to a
new job, you must be able to show that you have the skills, capabilities,

3052–8/95/0015$08.00/0 ©1995 American Chemical Society

knowledge, and personal characteristics expected for all chemists, as well as your own unique set of skills and capabilities.

Although all chemists possess a collection of skills, they frequently cannot recall examples quickly on demand, for example, under the stress of an interview. Recalling even significant contributions can be difficult in the anxiety of an interview or if called upon to discuss a résumé on the spot. To be prepared for rapid recall, and to avoid the embarrassment and confusion sometimes associated with nervousness, we recommend that you compile a fairly comprehensive list of your skills and accomplishments. This list will give you confidence and security during an interview; help you match your capabilities with the needs of the marketplace; help you prepare your résumé; and help you market yourself.

CREATING A LIST OF SKILLS AND ACCOMPLISHMENTS

Develop your own list of skills and accomplishments. Plan to take several days for the project. Sit down in the quiet of a room—at home or in a library—and mentally separate your life into segments, such as childhood and grade school, high school, college, graduate school, workplace and jobs held there, professional activities, community activities, hobbies, and sports. Think about your skills and accomplishments. You will not be able to accomplish this in one sitting; do this in 45-minute segments. As you reflect on the segments of your life, write down your accomplishments and skills in outline form, using key words. Most people are amazed at how long their lists become. Don't be concerned about whether an item is a skill or an accomplishment; you will sort these out later.

To be useful, your list must be sorted and simplified. Choose several major skills categories and then fit your accomplishments into these categories. To help you get started, refer to Figure 4. This list gives several categories of skills from which you might choose. Select the strongest and most significant areas where your skills and accomplishments lie, and use these categories. Keep your list up to date. All of these activities can be best accomplished using a spreadsheet package and a computer.

An example of a very abbreviated list of skills for one chemist is provided in Figure 5. This list was condensed from more than 275 skills and accomplishments and is provided for illustration only. Note that the cat-

Logic	Reasoning	Problem-Solving and Decision-Making
Analyzing data	Instrumentation	Reading and comprehension
Experimental design	Record-keeping	Decision-making and recommending
Organizational skills and time management	Communication (written or oral or computer)	Negotiation
Customer orientation and serving others	Language	Library skills

Figure 4. Potential categories of skills.

egories are largely different from those in Figure 4. This difference helps illustrate the fact that the categories chosen for any list are arbitrary and personal and depend on the intended use. If this list were to be used to apply for a specific job, it would need to be more focused. That is, some categories would be enlarged and others dropped. However, any such list can be used to generate a skill-based résumé. This list also provides a ready reference of examples for you to use and discuss in an interview.

Both the résumé and list of skills and accomplishments are dynamic. They are never complete, and they should focus on and vary with the need for different job applications. They evolve with the events of life. They should be maintained on a computer or word processor, and they should be continually revised and updated.

VALUES

Fifty-four candidates submitted résumés to Generic Chemical; five candidates were selected for interviews. Of the five persons selected, three will probably become the top candidates to compete for one job vacancy. From those three, one will receive an offer.

One of the factors most likely contributing to the job offer is the expected "fit" of the candidate in the corporate culture. This section will not deal with the overall interview process, but only one part of it, the role that values play in fitting into the corporate culture. To understand

Category of Skills	*Accomplishment*
Chemistry	Synthesized 2,2'-bipyrimidine
Chemistry	Synthesized other new pyrimidines and similar organic compounds
Chemistry	Developed new high-pressure pH titrator
Chemistry	Solved polymer alloy (vs. graft) problem
Chemistry	Coauthored definitive book on polymer analysis
Computer	Use Macintosh with Microsoft Word, Excel, Cricket Graph
Communication	Presented many papers at scientific meetings
Communication	Published 18 papers
Communication	Published several standard methods in ASTM book of standards
Communication	Wrote many internal research and administrative reviews within my company
Organizer	Established several new committees in ACS Division of Analytical Chemistry
Organizer	Ran technical program for Division of Analytical Chemistry for two ACS National Meetings
Organizer	Started luncheon group at work to discuss new and emerging technologies
Organizer	Started Canvassing Committee for EAS
Organizer	Award as best large section, ACS, during my chairmanship
Administrator	Managed analytical division in major company
Administrator	Organized and directed a major meeting at my company
Administrator	Ran ASTM Committee on Publications; Chairman STPs
Administrator	Helped our local technician affiliate group to become permanent organization
Administrator	Helped get approvals for ACS Division of Chemical Technicians

Figure 5. Example of an abbreviated list of skills and accomplishments.

the importance that fit plays in the hiring process, let's first try to understand the term corporate culture. As culture relates to organizations, it typically grows out of the operating beliefs and thinking of the chief executive officer and key management personnel. It then cascades downward through the rest of the organization. If key players in the organization change, especially the chief executive officer, the corporate culture may also change.

Organizations have different cultures because of the ideas and strategies that are used to manage them. One organization may emphasize entrepreneurial skills, whereas another may value traditional methods more highly. Neither is necessarily right or wrong. The culture derives from beliefs held by the key management at a particular point in time. Employees may be rewarded, disciplined, ignored, or shunted aside, depending on how their behaviors fit into the present organizational culture.

Value Matching

The importance of technical ability should not be understated in relation to job acquisition and performance. Without technical competence you could not carry out your job responsibilities effectively, but it is not the only factor on which you are evaluated or hired. Behaviors on the job, which directly relate to your overall job performance, are also considered. This area is where your personal beliefs and principles, your values, play an important part.

When you seek employment, therefore, it is important to have personal beliefs and ideas in harmony with those of a prospective employer. Unfortunately many people overlook the importance that value matching has for career success. Personal lack of harmony with the corporate culture may manifest itself as conflict in the workplace, and when there is conflict, you could be disciplined in some way. This conflict could lead to a smaller than expected salary increase, transfer to another job, or even job termination.

Some examples of conflict that may result from mismatched values are

- personality conflicts with supervisors;
- conflicts occurring in a work group if one employee attempts to innovate change in programs, activities, or methods when the unit prefers adherence to traditional practices;

- tension, which may develop in a work group if an employee seeks upward mobility, recognition, and power so vigorously that he or she causes serious damage to working relationships with other employees or managers;

- conflicts resulting between an employee and his or her manager over the need for feedback about work results, particularly if the employee demonstrates a need to work independently; and

- conflicts in an organization that rewards long work hours and extraordinary work commitment, especially if the employee works a normal eight-hour day on a continuing basis and provides highly productive results.

When these types of conflicts arise in the workplace, generally neither the organization nor the employee is satisfied.

Understanding Professional Values

To avoid the conflicts described above, we recommend that you identify your dominant personal values and the needs such values represent. This will help you to make your own determination about fit in an organization. Although there are many ways of examining values, the following values categories, with their definitions, can help you make a judgment about yourself and then about an organization's ability to support and nurture your strongest beliefs and principles.

ADVANCEMENT. Individuals tuned to advancement seek recognition in the workplace for their accomplishments and talents. Although often the recognition sought is monetary, it does not necessarily have to be so; for example, it might be a simple "thank you," an article about the accomplishment in an in-house newsletter, or a new job title.

Although most employees hope their careers will provide opportunities for upward mobility or developmental growth, the person driven by the need for advancement feels this drive very deeply. Such individuals, if not developing and advancing in their careers, will move to another organization where they can achieve these goals. Others will compromise—perhaps even plateau in their careers—if their primary value is being met in the current organization.

AUTONOMY. Individuals tuned to autonomy have a strong need to do things their way, at their own pace, and in line with their own standards. Because of the desire for a loose rein on their work activities, these individuals may find organizational life too restrictive and move into a work environment that offers much more freedom. This type of work environment can often be found in consulting and teaching.

BALANCE. Individuals seeking balance try to find equilibrium in all aspects of their lives. They do not want to choose between important family considerations, their career objectives, and their own self-development goals. Their lifestyle will influence decisions about relocations, family needs, work hours, and employee benefits.

CHALLENGE. Challenged individuals like competition, and their competitive nature drives them to overcome difficult obstacles and solve difficult problems. It is not unusual for such individuals to consider salary, job titles, and work area as secondary to the challenge of the task. When these individuals are employed in an organization that does not offer challenging work assignments, the challenge-driven individual can become bored or irritated with job assignments and with co-workers.

SECURITY. Security-driven individuals have a feeling that they must belong. They attempt to find employers who have a reputation for not reducing their work force. These individuals may experience a sense of personal failure because they are too content with the status quo and lack the ambition to pursue their career goals.

In today's workplace, individuals tuned to security may find it increasingly difficult to locate an organization that can provide the assurance of long-term employment. The security needed by these individuals must now come from a source other than the organization. Individuals must provide security for themselves. This type of security can be achieved if individuals keep their knowledge, skills, and abilities current with workplace needs.

Identifying Your Values

To identify your current one or two highest values, take 20–30 minutes when you are not hurried and are able to sit quietly and uninterrupted. Think back over your work experiences, your behaviors in various workplaces, and recall when you felt most fulfilled and content in your job. Also try to recall those experiences that seemed to bring you the most successes or conflicts at work. If you experienced conflicts in the workplace, do some hard self-assessment about the actions you took and why you took them. Think honestly about the part you played in the conflict and what you contributed to the conflict. To the extent that it is possible, strive to avoid earlier pitfalls and problems.

The values that are most important to you now will probably change in priority with time, but they do not change easily. For the most part, values change following strong emotional events that cause us to rethink what is important in our lives. Some examples are divorce, job loss, major geographical relocation, death of a family member, and birth of a baby. When a strong emotional event affects you deeply, it will be important for you to reassess your values to determine if your current employer's culture is able to match your new needs. If not, it is time to consider whether you should look for another job.

If you do undertake a job search, be sure to address some of these values and fit issues at your interview. Raise questions with the interviewer to help you evaluate your fit in the organization. Such questions may include questions about the organization's structure; determining what behaviors are rewarded in the culture; asking for the profile of a successful manager or technical person in the organization; and determining why the position is vacant. You also should observe the work environment and the personalities of the people you meet there. These data, in addition to any information that you gather through networking, will prove vital to you in the assessment of your fit.

COMMUNICATION

Good communication has always been a skill valued by employers, but in today's workplace, communication has taken on increased significance. There is hardly a job that does not require proficiency in oral or written communication, or both. Those bench chemists who in the past were able to work very independently of others are now finding that they can no longer do so. Chemical employers in the 1990s place a high premium on the ability to work in teams and to handle customer service.

Proficiency in both oral and written communication undergirds most of the chemical specialties covered in this book. Communication is so important to career transitions that it would be very difficult to make a change successfully without the competence to communicate well.

A Framework for Communication

Because of the importance of good communication, we present a framework that will help you to improve your overall performance. When this framework is understood and used properly, it can be a first step to improving interaction with others. It must, however, be reinforced by other communication training and practiced regularly to achieve noticeable results.

The framework uses a personal-style communication system called I-SPEAK Your Language, (I-SPEAK) developed by Drake Beam Morin, Inc., a leader in career transition consulting. I-SPEAK has its foundation in the work of the eminent Swiss psychologist Carl Jung, in particular his theory on the communication styles people use in their approaches to work and life.[1]

I-SPEAK Your Language helps identify not only your own primary communication style and your communication style under stress, but also the styles used by colleagues and management. It teaches you how to identify and modify your own style to work best with others in a variety of situations. There is no acceptable or unacceptable style. Developing your skill in recognizing and using these styles helps to establish good communication and does not necessarily mean changing your own style. By learning to identify various communication styles, you learn to speak the language of those styles. The four styles are described in Table I.

Description of Communication Styles

Each of the four styles associated with I-SPEAK Your Language has a time frame associated with it. The Intuitor focuses on the *future*; the Thinker moves between the *past, present, and future;* the Feeler looks back at the *past*; and the Senser refers to the *present*. It is possible to recognize a person's preferred communication style by listening for messages that are delivered regularly in the preferred time frame of the communicator. An Intuitor uses such words or phrases as "future trends," "long-term considerations," "outlook," and "prospects." The Thinker gives a step-by-step presentation of data or events and introduces a topic or idea in some logical order. The Feeler, who relates to

---

[1]I-SPEAK Your Language and I-SPEAK are registered trademarks of Harcourt Brace & Company. Copyright 1994 Drake Beam Morin, Inc. All rights reserved.

Table I. Characteristics Associated with Communication Styles

Style	Effective Application	Ineffective Application
Intuitor	Original	Unrealistic
	Imaginative	"Far-out"
	Creative	Fantasy-bound
	Broad-gauged	Scattered
	Charismatic	Devious
	Idealistic	Out-of-touch
	Intellectually tenacious	Dogmatic
	Ideological	Impractical
Thinker	Effective communicator	Verbose
	Deliberative	Indecisive
	Prudent	Overcautious
	Weighs alternatives	Overanalyzing
	Stabilizing	Unemotional
	Objective	Nondynamic
	Rational	Controlled and controlling
	Analytical	Overserious, rigid
Feeler	Spontaneous	Impulsive
	Persuasive	Manipulative
	Empathetic	Overpersonalizes
	Grasps traditional values	Sentimental
	Probing	Postponing
	Introspective	Guilt-ridden
	Draws out feelings of others	Stirs up conflict
	Loyal	Subjective
Senser	Pragmatic	Doesn't see long-range
	Assertive, directional	Status-seeking, self-involved
	Results-oriented	Acts first, then thinks
	Technically skilled	Lacks trust in others
	Objective; bases opinions on what he or she actually sees	Nit-picking
	Perfection seeking	

the past, often talks about "how things used to be." The Senser refers to what is important today and expects a quick turnaround of results.

From the brief overview of the various styles in Table I, you should be able to identify your particular style and those of others of interest to you. You may identify one or two styles that dominate your behavior, both at home and at work, when everything is operating normally. Then consider how your style might change when there is stress at home or at work. A change in your communication style is very likely to occur when you are under stress.

Next, think about the styles of some of your peers (perhaps team members) and of your supervisors and how their styles may differ from yours. If you have had a recent conflict or strong difference of opinion, either at home or at work, try to recall the events and reflect upon the style differences that may have caused miscommunication. At the next opportunity, try interacting with that person to confirm your opinion. When the style of communication is known, you will be able to consider ways of approaching that individual differently to discover a means of improving the relationship.

Here are expanded descriptions of the communication styles. You can use them to reinforce the patterns associated with each style.

THE INTUITOR. The Intuitor is often recognized as a global, conceptual thinker, someone who likes to see the "big picture." With this person there can, at times, be too much emphasis on the big picture while necessary technical details may be overlooked. The visionary capability of this person can frequently be found in upper management, where creative thinking is an important component of the position, or in chemical specialties that require creative work such as R&D. The high-intensity side of the Intuitor inclines this individual to stay predominantly in the pattern of future thinking; when recommending new programs or activities the Intuitor may appear to others as impractical and not sufficiently oriented to the immediate or short-term needs of the management or the organization.

THE THINKER. The Thinker prefers step-by-step logical thinking and uses the past, the present, and the future as a forum for doing analysis. Thinkers are analytical and objective and often are good communicators. They are very orderly, not only in their thinking but in the way in which they organize and maintain their work and home environments. Many individuals in the scientific community, who are trained to use problem-solving and reasoning skills to arrive at solutions, prefer this style of communicating with others. If the Thinker is in a high-intensity mode, he or she can become very talkative, giving too much detail to those who do not need it or desire it. The Thinker can be viewed as too rigid and impersonal or can be seen as slow to make decisions, as a result of being too cautious and overanalyzing data.

THE FEELER. The Feeler is a person who relates to past events. Feelers are known for their friendliness and empathy, in contrast with the Thinker, who prefers an objective, almost impersonal review of events. During a time of downsizing, the Feeler is among the first to consider personal issues surrounding a reduction in force, such as pay loss, family disruptions, and inequitable treatment. The Thinker, on the other hand, reviews the data to determine if procedure has been followed correctly, examines information about terminees' length of service, and so on. Thinkers care about people, but their outward view to the world is focused on the data associated with people, rather than the problems of people associated with the change. In a high-intensity mode, the Feeler can become very sentimental, act impulsively, and think too subjectively about events.

THE SENSER. The Senser lives in the "here and now," the present. The Senser is energized by seeking quick results and avoids lengthy and unnecessary discussions. Such individuals are quite accustomed to working on many projects at one time and for this reason want information presented in a concise format. If the Senser is a high-intensity user of this style, his or her interactions with you may seem like interrogations; the Senser may lack trust in others and nit-pick on details. The Senser may also "shoot from the hip" in reaction to events and not think before acting. Sensers are known for imposing their drive and their commitment to success on others, which at times can cause their co-workers or subordinates to feel overwhelmed by the pressure.

Sensers are frequently seen as the driving force within an organization. They are idea-oriented but won't respect the ideas of others unless they can see a practical and workable result. Sensers are such committed workers that they will go to any lengths to make a project successful.

Learning To Respond

Experience with the I-SPEAK Your Language system has shown that many chemists and chemical engineers prefer to use the Thinker and Senser patterns of communicating. Intuitors, although they have a presence in the scientific community, are usually in positions requiring creativity. Feelers are not often found in lab jobs; they may be found more frequently in specialties that require greater emphasis on people interactions such as human resources, certain units of marketing, and public relations.

Identifying a communication style is not the whole solution to improved communication. The next step is learning how to respond. Improved oral communication rests on the ability to recognize a style and to deliver a response in the manner that the listener prefers to use. By listening closely, you can begin to paraphrase some of the language used by another to communicate understanding. You can also improve communication by listening for clues about the types of issues that concern others and by incorporating them in your response. For example, Thinkers are known for objectivity and analytical thinking. In using this style they may unconsciously appear impersonal and cold in situations involving people. If the Thinker knows that he or she is dealing with a Feeler, the Thinker will be careful in an analysis to show effects upon people and the workplace. For example, an R&D manager who is asked to reduce staff size and to work with the human resources management to accomplish this should include as many aspects of people management as possible in making recommendations.

When drafting written communications, it is equally important to consider the time frame of the receiver and what he or she consider important issues. Intuitors, because they are interested in concepts, do not expect a great amount of detail. Intuitors are most interested in documents that are creative, visionary, and take into account long-term events that could influence outcomes. Thinkers, by contrast, expect a step-by-step discussion as to how results are to be achieved.

As explained earlier, the Feeler expects information to reflect the effects on people and wants to see a comparison of the past with what is being recommended for today or in the future. The Senser has a short attention span for reading and digesting information and usually wants a concise document.

By reflecting on your communication style, you can begin to see the strengths and weaknesses of your style and understand others' styles. What is more important, you can apply this knowledge in a number of ways:

- to improve your interviewing techniques,
- to enhance your work team,
- to negotiate more effectively, and
- to produce better presentations and written reports.

ଚ୦ ଜ୍ଞ

CHAPTER 4

SALARIES AND TRENDS IN THE CHEMICAL INDUSTRY

For many workers, a career transition means an increase in salary. For example, you might go from a bench career in which your salary has peaked to a management or sales position in the same firm that offers further salary growth. For some workers, however, a career transition may mean, at least in the short run, a salary cut. For example, if you move from a research position in a large pharmaceutical firm to a similar position in a contract research firm, your salary could drop substantially. The list "Sources of Current Salary Information" may be helpful.

FACTORS THAT INFLUENCE SALARIES

To get an idea of how your salary might be affected by a career move, you need to know what kinds of factors influence salaries. The largest influences on salaries are type of degree (bachelor's, master's, or doctorate), type of employer (academic or nonacademic), and length of experience.

NOTES: This chapter was written by Joan S. Burrelli, Senior Research Analyst in the ACS Department of Career Services.

Salaries generally have increased over time about as fast as the inflation rate. To update these salaries, inflate them by about 3–4% for each year beyond the publication date of this report.

SOURCES OF CURRENT SALARY INFORMATION

For current salary information, you may want to consult one or more of the following sources:

- Results of the annual ACS Salary Survey are published in *Chemical & Engineering News*, usually in the second issue in July. More detailed information on individual salaries can be obtained by calling the ACS Department of Career Services, (800) 227-5558.

- The American Association for University Professors (AAUP) and the College and University Personnel Association (CUPA) both conduct annual salary surveys of faculty and administrators. The results are usually reported in *The Chronicle of Higher Education*.

- The College Placement Council (CPC) reports starting salaries of new graduates in most disciplines in its quarterly *CPC Salary Survey*.

- *The Scientist* frequently reports salary survey results from a wide variety of organizations. It regularly reports on salaries in biotechnology, government, and toxicology.

- Sources such as *The Wall Street Journal* and *ChemicalWeek* report annually on executive pay in the chemical industry.

- The ACS Division of Chemical Information conducts a salary survey of people working in the field of chemical information services.

- *Working Woman* magazine annually publishes an article reporting results of salary surveys from a wide variety of sources. Salary information is published for a series of occupations, including chemistry.

- Robert Half International of Washington, DC, annually publishes its *Salary Guide,* which reports salaries in accounting, finance, and information systems.

- Other professional associations conducting salary surveys include
  The American College of Toxicology
  American Geologic Institute
  American Institute of Chemical Engineers
  American Mathematical Society
  Engineering Workforce Commission of the American Association of Engineering Societies
  Institute of Electrical and Electronic Engineers

Type of Degree

Those who obtain a Ph.D. in chemistry earn considerably more than those with a bachelor's or master's degree. Ph.D. chemists earn 50% more than B.S. or M.S. chemists, which translates to about $20,000 more at any given point in a career. It takes about 15 to 20 years of a career for bachelor's or master's chemists to earn the starting salary for a Ph.D. chemist. Obtaining a master's degree, however, does little in terms of increasing salary. M.S. chemists earn only slightly more (between $2000 and $5000) than bachelor's chemists.

Type of Employer

Industrial salaries range from the mid-$20,000s for beginning bachelor's or master's degree chemists to the mid-$60,000s for more experienced chemists. Beginning Ph.D.s in industry earn around $50,000, which increases to the mid-$80,000s for more experienced chemists. Chemists who are employed in industry generally earn more than those who are employed in government, and within industry, salaries are generally higher for those working in the petroleum or electronics industries.

Academic salaries are, on average, about 20% less than industrial salaries. Assistant professors typically earn between $32,000 and $40,000 for a nine-month contract. Associate professors earn between $37,000 and $48,000, and full professors earn between $48,000 and $68,000 on nine-month contracts. Academic salaries are generally slightly higher for those employed in public institutions, and from $7,000 to $14,000 higher for those employed in Ph.D.-granting schools.

High school chemistry teachers' salaries range from about $25,000 to about $40,000. Salaries in private schools are 20–30% lower than those in public schools. Salaries are also affected by the size of the school. Larger schools (those with 500 or more students) pay about $6,000 per year more than the average, and smaller schools (those with fewer than 50 students) pay about $6,000 per year less than the average.

Length of Experience

The length of experience in a particular career also has a strong influence on salaries. Salaries for most chemists rise fairly steadily for the first 20–25 years before plateauing. If you go from a position in which you have 15 years of experience to one in which you have no experience and expect to earn a starting salary, your salary might drop by as much as one-third.

Other Influences on Salary

Other factors influencing nonacademic salaries are number of subordinates and organizational size. Salaries increase with increased responsibilities. Chemists with bachelor's and master's degrees generally take on supervisory responsibilities within the first 5–9 years of their careers. For those who do so, salaries continue to rise, but for those who remain in nonsupervisory positions, salaries begin to plateau around 10–15 years into their careers. Most Ph.D.s start out in supervisory positions, and many take on more subordinates about five years after their Ph.D.s. After about 20 years, Ph.D.s in management positions earn about $15,000 more than those who remained in nonsupervisory research positions.

Organizational size also influences nonacademic salaries. Larger organizations (those with 2500 or more employees) generally pay about $5,000 to $8,000 more than smaller ones (those with fewer than 2500 employees).

Field of degree or field of work (e.g., organic chemistry, physical chemistry, polymer chemistry, analytical chemistry) generally makes little difference in salaries.

SALARIES IN NONTRADITIONAL CAREERS

Marketing and Sales

Salaries for experienced chemists with bachelor's or master's degrees in chemical marketing and sales are generally about $5,000 higher than those in research positions, but are not as high as those in management. Salaries in this area range from $40,000 to about $76,000.

Human Resources

Like salaries in general, salaries in human resources vary widely depending on factors such as organization size, geographic location, and length of experience. Some general guidelines, however, are that entry-level people in human resources earn salaries in the mid-$20,000s, and more senior staff earn around $50,000. Workers with advanced degrees may earn about $4,000 more. Human resources executives earn slightly over $100,000 in smaller firms and close to $130,000 in larger firms.

Law

Salaries in patent law are generally higher than those in chemistry—about 6% higher for Ph.D.s and about 30% higher for B.S. chemists. Ph.D. salaries in patent law range from about $65,000 to $90,000. Salaries for B.S. chemists range from $58,000 to $70,000.

Computers in Chemistry

Salaries in the computer field are also generally higher than those in chemistry—about 3% higher for Ph.D.s and about 30% higher for B.S. chemists. For Ph.D.s working in computer science, salaries range from $57,000 to $72,000. For B.S. chemists working in computer science, salaries range from $41,000 to $61,000. Starting salaries for computer programmers are generally in the mid-$20,000s, those for programmer analysts or systems analysts are generally in the mid-$30,000s, those for managers and software engineers are in the mid-$50,000s, and those for management information systems directors are generally in the mid-$60,000s. Salaries in large installations pay substantially (up to 30%) more than small installations.

Chemical Information Services

Salaries in the chemical information services vary greatly by the type of employer. Those employed in colleges or universities earn much less on average than those employed in industry. According to a survey by the ACS Chemical Information Division, salaries in industry are on average 35% more than those in academe. Most workers employed in this field have master's degrees. In industry, M.S. chemists earn about the same in the chemical information services as they would in R&D (from $45,000 to about $60,000). B.S. chemists in this field, however, earn much more ($40,000 to about $60,000) than they would in R&D.

Consulting

The most frequently charged hourly consulting rate for chemists is $100. Ph.D. chemists usually receive higher rates than B.S. or M.S. chemists.

The median hourly rate charged by B.S. chemists is $60; M.S. chemists generally charge $80; and Ph.D. chemists charge $100 per hour. The amount of consulting fee depends on the value of the consultant's expertise, overhead, and profit. Some consultants charge a daily rate rather than an hourly rate, and some charge a fixed fee for the service. To determine your hourly consulting rate, you might translate your current or former salary into an hourly rate and add 30% for fringe benefits. Your overhead and marketing costs should also be added into this. Other factors to consider when setting your consulting rate are length of job (the rate for longer jobs is less per day than that for shorter jobs) and size of firm (large firms pay more than small firms). You may charge half of the daily rate for travel days. In setting your fee as an expert witness in a trial, you may want to add 25–50% to your daily fee.

TRENDS IN THE CHEMICAL INDUSTRY

The nature of chemistry employment is changing as employers are changing the way they do business. Your decision to make a career transition may be influenced by one or more of these trends. Some of the more pervasive trends in industry are globalization, flattening of management, the changing nature of industrial competition, greater emphasis on specialists and experience, a blurring of traditional boundaries, and outsourcing of research and development.

Globalization

Increased competition from foreign companies increases the pressure to downsize and results in continuing expansion of U.S. R&D abroad. This means that those working for, or interested in working for, companies that are globally competitive should acquire foreign language skills and some interest in and appreciation of other cultures. For upward mobility within these organizations, it may be necessary to spend 3–5 years overseas.

Flattening of Management

As the business environment has become more competitive, companies have eliminated excess layers of management. This leanness results in less opportunity for traditional upward mobility. Technical ladders are becoming more prevalent as a means of advancing chemists. Because there are fewer layers of management, employees are expected to assume more individual responsibility for their work, are expected to have leadership qualities, and to some degree are expected to be risk takers.

Changing Nature of Industrial Competition

The nature of industrial competition is changing to an emphasis on continuous product improvement rather than on innovation. This change results in a greater business focus on providing service, reducing costs, and improving productivity. Employers look for people with some business training and people who can understand the commercial application of their research. Companies are focusing more on applied research and less on basic research, resulting in fewer opportunities in basic research and more in applied research.

Greater Emphasis on Specialists and Experience

In the past, companies hired generalists, who could then be trained for the company's needs. Increasingly, the emphasis is on hiring the specialist who is flexible. Companies want to hire specialists who can "hit the ground running," but the employee must be broadly trained enough to be able to change directions later as industry's needs change. Because of increased competition and a greater applied or customer orientation, companies are looking for practical experience from prospective employees.

Blurring of Traditional Boundaries

Rather than a hierarchical reporting structure and separate functional departments and divisions, companies are increasingly moving toward multidisciplinary teams in "seamless" organizations. Departments and other organizational units are more interdependent, and traditional boundaries between disciplines are becoming less distinct. The blurring of traditional boundaries implies that more opportunities will be available at the "fringes" of chemistry, i.e., at the intersections of chemistry and other disciplines. It also means a greater need for interpersonal skills, for task orientation, and for greater flexibility, i.e., more emphasis on problem-solving skills. Because of this development, employees need to develop their problem-solving skills, their interpersonal skills, their leadership skills, and their communication skills.

Outsourcing of Research and Development

Large companies are contracting out much of the R&D that they formerly conducted in-house. More research alliances are being established with universities, private labs, research institutes, and other partners. Thus, employment opportunities will increase in small or medium-sized companies and will decrease in large companies.

୫୦ ୯୪
CHAPTER 5

PERSONAL DATA FORMATS

This chapter is deliberately titled *Personal Data Formats* to avoid nonspecific use of the term "résumé." When making a job or career transition, the work environment being considered governs the format to be selected. A résumé is not appropriate for every workplace. Chemists work predominantly in three segments of the economy: industry, academe, and the federal government. The format for each of these segments, the manner in which the data should be presented, and the selection of data to be highlighted vary accordingly.

Generally speaking, the formats to use in each of the three work areas are as presented in Table II. A discussion of each format follows.

Regardless of which personal data format is used, its goal or purpose is to secure an interview. Thus it is not meant to be a lengthy autobiography. Much of the information needed from you will be developed in the interview. Your résumé is intended to invite interest by highlighting information that you believe will be a possible match with the employer's needs and critical for effective performance in the position.

The curriculum vitae (CV) and the government application forms require more attention to detail than a résumé, and work experience is presented chronologically. Select oral presentations and publications that demonstrate your research interests and that will help you to strengthen your CV. For the government forms, carefully develop the narrative section outlining your chronological work history, and prepare a supplemen-

3052–8/95/0037$08.00/0 ©1995 American Chemical Society

Table II. Personal Data Formats

Work Area	Format	Exceptions
Industry	résumé	In some small biotech firms, a curriculum vitae is preferred because of the academic environment of the workplace. When considering work in this community, inquire about the type of form preferred.
Academe	curriculum vitae	
Federal government	SF-171	
State and local governments	application form	In some instances, résumés will be accepted for an initial interview.

tal statement that will give you the opportunity to emphasize your accomplishments.

PROVIDING YOUR CREDENTIALS

Chemists serve in a professional capacity in our society. You are hired as a professional and you fill the job as a professional. Thus, from time to time you will be asked to provide your credentials as a chemist. This request is made especially with new hires, or if you are changing a career and moving from one organization to another, or if you are in the business of consulting. The significance and impact of your credentials will vary with their content, with the job you are seeking to fill, and with the organization that is hiring you. Sometimes you will want to list or mention all of your credentials; sometimes you will be more selective. What you mention will vary with the circumstances. Some or all of this information is used on your résumé and in your interview (see box).

Figure 6 is a framework for credentials for a chemist looking for a research position in a large company. Generally the most significant information is given first, beginning with the advanced degree. You can use this framework and adjust it to your own circumstances (experience, job sought) for your personal use.

WHAT A RÉSUMÉ SHOULD PROVIDE

A Sense of Purpose

Give the reader an indication of the type of position you seek. The reader should not reach his or her own conclusions about your job goals and your qualifications for the opportunity because the wrong conclusions may be drawn.

Emphasis on Achievements

Include achievements as bulleted statements under a skills listing or, if using a chronological format, ensure that the narrative or bulleted statements reflect your achievements. Try to differentiate yourself from other applicants in a positive way (i.e., showing leadership skills by professional roles).

Accuracy and Credibility

Never exaggerate your qualifications. Be truthful and accurate. Do not give false information or inaccurate job titles. Present your qualifications in the most impressive light; however, misrepresentation can cost you an interview, even a job.

Clarity and Simplicity

Remove unnecessary words to facilitate scanning of the document.

An Attractive Package—Design And Layout

Leave a one-inch margin all around your résumé. The layout should be clean, with ample white space and should not exceed two pages. Pay attention to the type of font you use. Fonts like Prestige Elite, Courier, Times Roman, and Geneva print clearly. Use 12-point size type so that your résumé will be easy to review.

Salesmanship—Measurable Facts That Appeal

Try to develop a match between your personal assets and those desired by the organization to which you are applying.

A Sense of the Person Behind the Document

Your information about work and outside activities should convey a picture of you as a well-rounded person.

Educational Degree

- Ph.D., M.S., foreign equivalent. Cite the university, mentor or major professor, thesis title, and year.
- B.S., B.A., or A.B. Cite the university or college and year.
- A.A. Cite the university or college and year.

Job Experience

- Professional positions held. Cite institution names, locations, and dates.
- Professional contributions. Cite advancements, promotions, awards within organizations (without violating propriety).

Patents and Publications

In all cases, cite the literature using the professional format.
- Patents
- Original papers
- Books
- Chapters and reviews
- Monographs (data, summaries, critiques)

References

In all cases, give full professional names, addresses, and phone numbers.
- Professors
- Supervisors and managers
- Colleagues
- Community leaders

Presentations

In all cases, cite the examples using the professional format.
- Original papers
- College or university courses
- Workshops, seminars, conferences

Professional Activities

Cite the name of the organization, location, dates, and title of position.
- President, chairman, officer
- Committee chair, officer
- Member of task force
- Member of society

Additional Study

- University, college, technical school courses
- Workshops, seminars, conferences

Figure 6. A framework for credentials.

RÉSUMÉ FORMATS

There are as many résumé formats as there are books on the subject. The résumé formats that appear to be the most accepted by the scientific community are the chronological and combination résumés. Each format has its distinct advantages. Figures 7 and 8 are résumés.

The Chronological Résumé

The chronological résumé provides information on job history. It is intended to present the candidate's job titles, the levels of responsibility held in various positions, and the names of previous employers (Figure 7). It is most useful when a job candidate wants

- to show good career progression without work gaps,
- to stay in the same line of work,
- to highlight employment at particular firms, and
- to highlight the level of work activity.

The Combination Résumé

The combination résumé highlights skills, knowledge, and accomplishments rather than job history. The combination résumé is useful for more experienced chemists and when

- changing career paths,
- seeking lower level job responsibility than held in previous employment,
- avoiding emphasis on gaps in employment or on periods of unemployment,
- diverting attention away from age, and
- avoiding noting that your experience or education does not exactly match the stated or known job requirements.

Revising a Chronological Résumé for Transitions

We have included examples of two résumés: the original chronological résumé and a revised combination version (Figure 8). This person's initial job goal is to become a patent liaison and move eventually into pat-

JOHN J. CHEMIST
1234 Antimony Lane
Catalyst, OH 44133
(216) 555-9876 (H)
(216) 555-1234 (O)

SUMMARY

University of Michigan Ph.D. chemist with experience in management of R&D and strategic customer and quality orientations. Brings practical perspective to solving complex technical problems. Thirteen years of experience in specialty chemicals and coatings includes

- Project Management
- New Product Development
- Opportunity Assessment
- Technology Transfer
- Coating Cure Chemistries

- Symposium Organization
- Patent Tracking System
- R&D Cycle-Time Reduction
- Synthetic Organic Chemistry
- Polymer Chemistry

EMPLOYMENT

THE GLIDDEN COMPANY/ICI Paints North America, Cleveland, OH   **1990-1993**
Technical Manager, Polymer Chemistry
Managed six chemists (Ph.D. and B.S.) in a basic polymer synthesis group. Improved interaction between the research group and coatings application groups.

- Initiated reevaluation of new cure agent for general powder coatings. Led group that improved gloss and doubled flexibility.
- Organized and led international team to prepare recyclable release coating for PET film. Transferred technology to development group in England.
- Led and coordinated international team characterizing new resin for latex house paints to determine compliance with government regulations (PMN) on commercialization of new materials.
- Coordinated work with customer to develop in-mold coating for new reaction injection molding (RIM) process. Demonstrated feasibility of the approach, transferred to development group, and supported coating formulation.
- Initiated study that identified stable formulation for new resins for latex house paint. Results kept project on target for commercialization.
- Improved lab safety by managing efforts to reduce contact with hazardous materials, dispose of hazardous chemicals, and meet OSHA regulations.
- Coordinated science education program with local elementary school district to hold a Science Olympiad by teaming volunteers with science teachers.

THE SHERWIN WILLIAMS COMPANY, Chicago, IL   **1983-1990**
Senior Scientist/Group Leader
Polymer Chemistry, Automotive Aftermarket Division (1986-1990)
Supervised a group of 5-7 chemists (Ph.D. and B.S.) and technicians in development of polymers for new coating technologies to improve performance and competitive edge.

Figure 7. A chronological résumé.

- Led project to develop new automotive refinish coating. Supervised in-plant production of initial batch of resin to support commercialization.
- Initiated and implemented company-wide technical symposium that drew more than 90 people and more than 40 presentations. Improved communications between technical staff in all R&D centers.
- Developed and implemented a patent database. Tracked all ideas from initial disclosure through maintenance fee payment schedule. Provided management with data for developing patent strategy.

Research Chemist, Central Research Laboratory (1983-1986)
Designed and developed catalysts, curing agents, and performance-improving additives for solvent-based paints.
- Conceived and developed a blocked catalyst with a sharp activation temperature for low-bake coatings for appliances.

ETHYL CORPORATION, Detroit, MI   **1979-1983**
Research Chemist, Industrial Chemicals
Designed and developed synthesis of specialty chemicals, including agricultural and pharmaceutical intermediates and monomers.
- Persuaded management to fund project to adapt new process technology to increase specialty chemical business opportunities. Received U.S. Patent 4,490,564 on Mixed Solvent Recovery of 4,4'-Dihydroxybiphenyl.

EDUCATION
Ph.D.-Chemistry, University of Michigan, 1979
Thesis area: Regioselective synthesis of sugars

M.S.-Chemistry, University of Michigan, 1976

B.S.-Chemistry, University of Cincinnati, 1973
Senior thesis area: Bacteria growth inhibition study

PROFESSIONAL ACTIVITIES
Member American Chemical Society
Edison Polymer Innovation Corporation Representative
North Royalton Association for the Talented and Gifted
ACS High School Education Committee
Glidden Educational Task Force
Science Olympiad Coordinator

PUBLICATION
Regioselective Synthesis of Islandicine and Digitopurpone, T. A. Becket, J. J. Chemist, I. M. Good; *Journal of Organic Chemistry* (1980), 45, 516.

Figure 7. *Continued.*

JOHN J. CHEMIST
1234 Antimony Lane
Catalyst, OH 44133
(216) 555-9876 (H)
(216) 555-1234 (O)

SUMMARY:

Experienced chemist seeks position as Patent Liaison. Can bring to position good understanding of Opportunity Assessment and Technology Transfer, and Patent Tracking Systems. Excellent oral and written skills.

KNOWLEDGE AND SKILLS:

Opportunity Assessment and Technology Transfer

- Organized and led international team to prepare recyclable release coating for PET film.
- Interacted with members of international group to understand their needs and interpret applications of research.
- Led and coordinated international team characterizing new resin for latex house paints.
- Led development and in-plant production of new automotive refinish coating.
- Developed new reaction injection molding (RIM) process with customer.

Patent Tracking Systems

- Developed and implemented a patent database, which tracked all ideas from initial disclosure through maintenance fee payment schedule, providing management with data for developing patent strategy.
- Over a nine-month period worked part-time to build patent file, which provided incumbent with excellent understanding of the patent process.
- Discussed patent process needs with legal staff to determine type of data to gather; designed format for data collection to develop other needed information.
- Interacted with members of technical research staff and secured their trust in gathering necessary patent information to build data file.

Technical Skills in Chemistry

Organic and Polymer
- Initiated reevaluation of new cure agent for general powder coatings.
- Initiated study that identified stable formulation for new resins for latex house paint.
- Designed and developed catalysts, curing agents, performance-improving additives, and agricultural and pharmaceutical intermediates.

Communication Skills—Oral and Written
- Initiated and implemented company-wide technical symposium to improve communications between technical staff in all R&D centers.

Figure 8. The revised combination résumé.

- Worked with technical staff to improve lab safety by managing efforts to reduce contact with hazardous materials, dispose of hazardous chemicals, and meet OSHA regulations.
- Coordinated science education program with local elementary school district by teaming volunteers with science teachers.
- Drafted written recommendations for the implementation of a patent tracking system, which were accepted by management.
- After a thorough analysis, drafted a written report on the challenges of working with the talented and gifted for an Ohio school system.

EMPLOYMENT:

1990-1993	Technical Manager, Polymer Chemistry The Glidden Company, Cleveland, Ohio
1983-1990	Senior Scientist/Group Leader The Sherwin Williams Company
1979-1983	Research Chemist Ethyl Corporation

EDUCATION:

Ph.D., University of Michigan, 1979. "Regioselective Synthesis of Sugars," Professor T. A. Becket.

M.S., University of Michigan, Chemistry, 1976.

B.S., University of Cincinnati, 1973.

PROFESSIONAL ACTIVITIES:

American Chemical Society
Edison Polymer Innovation Corporation Representative
Catalyst, Ohio's Association for the Gifted and Talented
American Chemical Society: High School Education Committee
Paint Company Educational Task Force
Science Olympiad Coordinator

Figure 8. *Continued.*

ent law. We reworked the chronological résumé and developed the combination résumé to highlight certain skills and knowledge, which we know are important in the intellectual property work community, to strengthen this chemist's candidacy for a patent liaison position.

CURRICULUM VITAE

The curriculum vitae, often referred to as a CV, is the format most appropriate for academic employment. It is not as brief as a traditional résumé. A brief research statement, a publications list, and oral presentations are additionally important in preparing this document. The choices of information to include in the document rest on judgment about what will be important to the particular academic institution. It will take a certain amount of networking and investigating on your part to uncover the data.

Generally, in a four-year college or two-year community college more emphasis is given to teaching skills and working with students, whereas in a Ph.D.-granting institution, more emphasis may be placed upon the type of research to be brought to the university and how it will enhance its overall research effort.

Figure 9 shows an example of a CV; to save space, the example does not include a listing of books and publications, which should be provided with the CV.

SF-171

The SF-171 is the job application used by the federal government. It is a specific, detailed form and may take several hours to complete. A copy of the SF-171 is available from any government personnel office or the Office of Personnel Management (OPM). Points are assigned to each response; the rating represents the total of these points. The following tips may help you in the preparation of the SF-171.

Always complete the entire SF-171. You may inquire whether a résumé is acceptable for the initial screening. If a résumé is acceptable, use one of the formats that has already been described. In most situations, however, a résumé is not a substitute for the SF-171. When completing the SF-171, remember the following:

• Address the knowledge, skills, and abilities required for the position.

- Carefully complete each experience block you need to describe your work experience. Unless you qualify based on education alone, your rating will depend on your description of previous jobs. Do not leave out any jobs you held during the past 10 years.
- Review the job announcement carefully; address all requirements.
- If a supplemental statement is optional, write one.
- File the form by the deadline.
- Return the form by first class mail. Do not use certified or registered mail.
- The SF-171 is reviewed either by a panel or staffing specialist.
- The selecting official cannot be on the screening panel.
- Telephone the staffing specialist to determine if you are on the certification list.
- Veterans and disabled veterans receive 5 extra points.
- An SF-171 software package is available from most computer software retailers for approximately $40.

KEEPING A RECORD

At the conclusion of your job search we recommend that you keep all of the forms (résumés, CVs, or SF-171s) from your search in a file for several reasons. First, you may be called in the future about a résumé you submitted to a company that had no suitable vacancy at the time. Second, you should examine your résumés for their success ratio, that is, their ability to produce interviews in specific types of organizations. This analysis will help you to determine how to repackage yourself in future job searches. Third, you should review your résumés periodically to determine how your job goals may have changed and why.

For more help with résumé writing, consult the ACS publication *Tips on Résumé Preparation* or ask for assistance from an ACS Career Consultant. See the section on ACS Career Services in Appendix I (pp 158–162) for more information.

TED SANDERS
Professor and Head
Department of Chemistry
Eastern University
Boston, MA 02134
(617) 555-3060

EDUCATION:

Ph.D., 1966   Slippery Rock College, Thesis: The Chemistry of Balonium (Prof. I. M. Good)

B.S., 1960   University of Arizona

EXPERIENCE:

1985-   Head, Department of Chemistry
1979   Visiting Professor, Eth, Zurich
1977-   Professor of Chemistry
1972-77   Associate Professor
1966-72   Assistant Professor, Eastern University
1966   Research Associate, The Hospital for Special Surgery
1960-66   Teaching Assistant, Research Assistant, Instructor, University of Arizona

RESEARCH INTERESTS:

Solid-state chemistry; crystal growth. Crystal and structural chemistry of complex oxides. Solid-state chemistry of heterogeneous catalysts.

PROFESSIONAL ACTIVITIES:

Chair, Search Committee for Director of Eastern University Honors Program, 1992-
Member, Honors Program Review Committee, 1992
Member, Task Force on the Financing of the Council for Chemical Research, 1991
Review Panel, NSF Initiative for High-Temperature Materials, 1989
Member, Advisory Board, Science and Technology Institute of New England, 1988-90
President, Board of Directors, Science and Technology Institute of New England, 1987-88
Member, *ad hoc* Sub-Committee on Solid-State Chemistry, NRC/NAS, 1977
Chair, Research Council, Eastern University Research Foundation, 1976-77
Member, Eastern University Research Foundation, 1973-77
Member, Eastern University Patent Advisory Committee, 1973-77
Member, Institute of Materials Science, Eastern University, 1972-77
Director, NSF Undergraduate Research Participation Program in Chemistry at Eastern University, 1970 and 1971
Northeastern Section Representative, Northeast Region Steering Committee, American Chemical Society, 1969-72
Secretary-Treasurer, Northeastern Section, American Chemical Society, 1967-71

Figure 9. The curriculum vitae.

PROFESSIONAL SOCIETIES:

American Association for the Advancement of Science
American Association of University Professors
American Ceramic Society
American Chemical Society
American Crystallographic Association
Materials Research Society
Mineralogical Society of America
Phi Lambda Upsilon

Figure 9. *Continued.*

CHAPTER 6

NETWORKING

In planning a job transition, one of the most important activities is networking. Networking is not an activity that can be done at the last minute or begun when you receive a "pink slip." To be most helpful, networking should start at the beginning of your career and should continue throughout it. Not everyone realizes this fact, however, and for many people, networking is begun some time well into a career or at a time of crisis.

Networking is not asking everyone you know for a job. Rather, you are seeking the help of people to gather information, ideas, and introductions useful to you in your job search. Your network is, or will become, part of your market research. Research has shown that approximately 75% or more of all jobs are hidden; that is, they are not currently being advertised. These jobs are located by networking, in which they are pointed out by one of your network members. These hidden jobs also offer you less competition from fellow job seekers because they are not yet openly advertised.

Networking is the process of building up a group of friends and colleagues with whom you can interact on an immediate or casual basis to discuss common interests or challenges, or to help one another in time of need. Some people in your network will be close to you, others more loosely connected. They are all part of your network. You may associate with several networks in your life, each relating to an area of interest or activity, work, or hobby. Network members may include professors, mentors, colleagues, supervisors, subordinates, customers, clients, con-

3052–8/95/0051$08.00/0 ©1995 American Chemical Society

sultants, members of hobby groups, sports groups, relatives, neighbors, physicians, or your dentist or lawyer.

BUILDING A NETWORK

If you are in a position where you need advice about how to build a network, here are some tips:

- Attend scientific talks and introduce yourself to some of the speakers afterward. Show an interest in their work. Talk about the work with others attending the meeting.
- Join several professional societies and go to their meetings, both local and national.
- Offer to participate on committees or task groups, and when appointed, follow through.
- Join clubs in your areas of interest.
- Practice small talk whenever possible. Ask people to join you for lunch. Use your telephone to call colleagues and chat.
- Use your computer e-mail to keep up contacts, but don't eliminate personal and voice contacts.

USING THE NETWORK

First and foremost, you must tell members of your network that you are looking for a job and you would like their help. Some people worry about using others or about imposing on others. Don't let this get in the way, but don't become a "network pest" either. Your main concern is to obtain the necessary information to get an interview, to get a job, to secure employment. Your turn will come to pay back someone else in the network. Giving and receiving are part of your network objectives. Whether you are giving or receiving at any given time is simply a matter of need and opportunity. While you are building or using your network, try the following:

- Keep track of your network members.
- Generate a computer list or a card file with names, addresses, and phone numbers.
- Jot down the date, time, and subject of conversations.
- Hand out business cards.

- Tell people about your current activities and job search.
- Be assertive but, of course, be appropriate and polite.
- Follow up on conversations.
- Be open to and follow up leads and suggestions that are given to you.
- Participate, and above all, initiate, *initiate*, **initiate** communication.

At all times during your networking process and job search you should be ready for immediate follow-up. Have a résumé ready to submit and be prepared for an interview.

Once you have found a job, let the people in your network know. Thank them for their help during your search. Also, remember that networking is not a one-time activity. Maintain your network; chances are you may need it again.

❧ ❧
CHAPTER 7

INTERVIEWING

All of the preparations that you make—developing a list of skills and accomplishments, establishing a network and maintaining the contacts that you make through it, and preparing a résumé—are done in order to obtain a job interview. Once you have arrived at the interview stage you will need to present your credentials in the best possible light. Only a small number of contacts and résumés lead to an interview, and only a few interviews lead to a job offer. Your efforts early in the process will ensure a more successful outcome. There are several types of interviews, but most of these are either screening or site interviews.

THE SCREENING INTERVIEW

On-campus interviews, employment clearinghouse interviews, and telephone calls from recruiters to candidates for more information are examples of screening interviews. Screening interviews are usually short and are used to determine if there is interest in bringing you in for a second, more in-depth interview. The interviewer typically tries to determine if you have the required technical skills, to assess your accomplishments with respect to the organization's needs, and to make an initial assessment of your fit in the organizational culture.

Although at first glance it may appear as if the screening interview is only for the benefit of the prospective organization, you should also address these issues to determine if you are sufficiently interested in proceeding to a site interview.

3052–8/95/0055$08.00/0 ©1995 American Chemical Society

THE SITE INTERVIEW

The site interview is arranged for the organization to determine whether or not to make a job offer and, for you, whether or not to accept an offer if one is made. Discussions are held in much greater depth, and each side is much more exposed to the other, so that rational and wise decisions can be made. Issues typically addressed by the organization are

- collective opinions of your technical capabilities and your compatibility with the organizational culture;

- your qualifications for a specific job, i.e., your technical knowledge, skill, research interests, and intellectual strength; and

- your ability to contribute and to grow in the organization.

You should assess the opportunity to contribute, grow, and enjoy working in the organization in a specific department and under management that will fit your needs. To prepare effectively for the site interview, you need to understand the process. The sequence may vary, but, with larger companies, you typically will

- meet with representatives of the Human Resources Department;

- meet with various technical people, including your prospective supervisor, to discuss technical problems and demonstrate your skills and abilities;

- be asked to give a technical presentation on your recent work (Ph.D. thesis if applicable, current work if employed);[1]

- be introduced to managers who will be especially interested in your values, beliefs, and potential fit in the organization;

- tour the laboratory, plant, or both; and

- review the company's employment policies and benefit programs.

You may receive an offer at the time of the site visit, or you will be told when to expect a decision. If you are not told, ask what the timetable is. If you do not hear by that time, call to follow up on the status of your application.

---

[1]You are not expected to disclose confidential or proprietary materials, and it is perfectly acceptable to present your research in general terms or to use nonproprietary examples. At the beginning of your presentation, mention that you plan to do so.

Preparing for the Interview

Whether you will participate in a screening interview or a site interview, there are a number of things that you should bear in mind. Other candidates will provide significant competition for job offers, and you do not want to eliminate yourself from consideration because you present yourself poorly. Therefore, when you have an interview appointment, do the following:

- Be punctual.
- Be alert and courteous, interested, curious, enthusiastic, and confident.
- Be flexible and adaptable.
- Show that you are strong as an individual, but are also a good team player.
- Present a good personal appearance.
- Stay away from extremes in all discussions.

Following Up After the Interview

After the interview, several decisions have to be made: a decision by the organization whether to hire you or not, when to make you an offer, and what salary to offer to you. You must decide whether to accept an offer or not, whether you should attempt to negotiate further (for example, for salary or level of responsibility), and when a reply is expected from you. It should be clear when these decisions will be made and how they will be communicated. Larger organizations, with established human resources offices, are aware of such issues, and normally they take the lead in structuring offers and further communications. Smaller organizations may not have a human resources office; you may need to ask more questions of the hiring manager if he or she provides insufficient information. If the organization does not give you a timetable as to when these decisions will be made, you should at least negotiate a decision time, express your interest in the position, and indicate that you will call to find out the status of your application.

INTERVIEWING TIPS

- Before you come to the interview, practice possible interview questions with a friend or spouse. Talk to the mirror or into a tape recorder if necessary to become accustomed to hearing your own voice.

- If possible, have a friend videotape your practice interview and critique it. ACS offers taped mock interview sessions at national meetings; take advantage of these. For research positions it is very likely that you will be asked to make a presentation of your current work. If there are complex issues to present, have hard copies of material available to hand out.

- Before the interview, learn as much as you can about the organization so that you will be in a position to ask intelligent questions. You can do this by reading the organization's annual report or news and financial articles about the organization.

- Before the interview, do some isometric exercises: Tense and relax your muscles, breathe deeply a few times, and use this time to observe your surroundings for clues about the organization's values.

- During the interview, be a good communicator. Try to do at least 50% of the talking. Listen to questions before answering.

- Show that you have good technical strength; be articulate about your skills and accomplishments.

- Note your interviewer's communication style and his or her demeanor (body language, tone of voice, tempo of discussion, facial expressions). To the extent possible, adjust your communication style to be similar to that of your interviewer, but do not mimic him or her.

- Maintain good eye contact with the interviewer and monitor his or her reaction to questions and answers.

- Know your own values and goals.

- Be prepared to discuss your credentials.

- Don't oversell or undersell yourself or your ideas.

- Be sure that you close your interview on a positive note.

- After the interview, evaluate your role objectively and be careful to avoid false readings of approval. Some interviewers are very friendly but hide their true feelings and evaluation.

ॐ ॐ
CHAPTER 8

MAKING A JOB DECISION

Whether you are on a long-term course, diligently preparing for future transitions from one job to another or from one career to another, or whether you are at a decision point about accepting a job offer, you should occasionally bring together into one place all of your job search information. The guidelines presented in this chapter suggest the steps necessary for you to make a good decision.

ORGANIZING YOUR DATA

The data you need to accomplish a successful job search and to make a successful job decision include many topics. You should keep a copy of each of the following:

- résumé you have used in your job search;
- copies of any presentations you have made;
- your list of publications and your list of references;
- your list of skills and accomplishments;
- your list of values;
- a copy (computer list or card file) of your network names, addresses, and telephone numbers;
- copies of published descriptions of jobs you have interviewed for or would like to have;
- a copy of the names, telephone numbers, and correspondence of people in companies where you have interviewed;

3052–8/95/0059$08.00/0 ©1995 American Chemical Society

- financial and benefit requirements; and
- your interview notes.

Assemble and review these materials to help you make the best decision about accepting a job offer or continuing on a particular career path. If you are in the planning stage of a transition, use either the sphere of opportunity (Figure 2) or the zero-sum line of opportunity (Figure 3) method outlined in Chapter 2 to help you envision how to continue. If you are at the point of a job offer and need to make a decision, use the rest of this chapter to aid you in making your decision.

ANALYZING YOUR DATA

During the decision-making process, review your correspondence and interview notes. Write down the skills and knowledge needed to do the job, the principal values of the particular organization, and your evaluation of the communications styles of your future bosses. Then answer these four questions:

- Do your skills and knowledge match those needed to do the job successfully?
- Are your professional values and those of the company compatible?
- Will you be able to communicate successfully with your future colleagues and management?
- Are the pay, benefits, and location acceptable?

If you answered all of these questions affirmatively, you will likely accept the job offer. If you answered "no" or "maybe" to one or more of the questions, you probably should seriously reconsider the offer.

Often, such analysis puts you in a gray area. There are some positives and some negatives, but it is not obvious how heavily they should be weighted, and it is difficult to tell just what should persuade you in your decision.

If your analysis falls into the gray area, there are two approaches you can take to aid you in making your decision. First, you should continue your analysis by looking at the problem from the "other side."

- What could go wrong if you do accept or reject the job?
- What negative consequences will there be?
- Identify the consequences for each case. Write them down. Think about them. Are they very important? Can they be circumvented or overcome?

Secondly, you should also consider alternatives to this job offer. Write these down. You may identify others, but here are several possibilities.

- You may have received other job offers. Where do they fall in your analysis?
- You can remain in your present job.
- You could go to or stay in school to further your education, in the same or a different discipline.
- You could take an interim job while continuing your search.

Whatever your alternatives, try to evaluate the consequences of following one or another of them rather than accepting a particular job offer. Would you choose one of these alternatives if you could? Why not? Throughout your analysis, consider your potential fit in the company and your fit where you are now. Keep in mind that skill or capability defects often can be overcome by training and education, but value differences or beliefs are quite difficult to overcome.

Now you are ready to make a final decision. Use your data and your analysis. What you must decide may be clear to you, or you may still be in a gray area. However, you have looked at the job offer quantitatively considering your skills, values, communication style, and the job's salary and benefits; you have looked at the offer from more than one perspective; you have considered many of the consequences of your decision; and you have looked at alternatives. Make a decision and implement a course of action. Telephone the company to accept or reject the job offer. Follow up in writing. You have done the best you could do.

Be happy. Celebrate. Be confident that you have made the right decision.

DETERMINING A MATCH

Matching the qualifications of a job to your personal capabilities is at the heart of the job transition process. To penetrate a "new" market, whether it is from one chemical bench specialty to another or to a nontraditional career, you must determine those qualifications that are most important in your chosen career path. Matching is achieved when you possess the knowledge and skills necessary to undertake the set of tasks expected by an employer.

This book discusses the importance for chemists of assessing one's knowledge, skills, and abilities because they give clues to career connections. Transitions, for the most part, are built upon small, incremental steps that the individual creates with time. Each step of newly acquired knowledge gives the individual an opportunity to enhance current skills or develop new ones.

Figure 1 (in Chapter 2) illustrating the core skills of chemists, is an invaluable tool to help you discover which of the core skills comprise your strengths and which ones need further development. The list of core skills should help to assist you in making connections between your background and those skills that are outlined in the career descriptions. When a connection is indicated, refer to Chapter 2 to help you assess your options.

Although the core skills of chemists are very evident in the traditional science career, they become less evident in nontraditional career jobs. However, this situation does not mean that chemistry is less important in nontraditional career jobs. The transition to a nontraditional career cannot be accomplished without maintaining competence in the chemistry knowledge that you possess. It is, therefore, important always to keep your basic knowledge of chemistry up to date, along with the additional skills that will be required in your new career path.

KEEPING YOURSELF MARKETABLE

Once you have made a successful transition, you need to keep yourself marketable. After all, job security is the ability to find another job. Following are several ways to keep yourself marketable:

- *Maintain your professional network.* Stay in touch with your former work colleagues and professional associates. Participate fully in local professional organizations, such as ACS local sections.

- *Expand your knowledge and skill base.* Keep current in your field. Stay current in what is generally happening in chemistry, as well as what is happening in your area of chemistry. Develop your communication skills, your interpersonal skills, and your ability to work on a team. Learn another language, such as German, Chinese, Russian, or Japanese. Learn or develop business skills. Expand or develop your skills and knowledge in related disciplines, e.g., physics, biology, or engineering.

- *Make yourself and your abilities very visible.* If you are interviewing, emphasize what makes you special. Take every opportunity you can to publish and present.

- *Most important, be flexible.* Keep your options open by relocating. Continue your education in a new or related field. Expand the responsibilities in your current job by making a lateral move, taking a short-term assignment, participating in task forces, or anything else that gives you a chance to see if the grass is greener elsewhere in the company.

80 C8
PART II

LABORATORY CAREERS

ଽଧ ଔଷ
CHAPTER 9

ANALYTICAL CHEMISTRY

Analytical chemistry is the science of obtaining, processing, and communicating information about the composition and structure of matter. It includes developing or transferring theory and practice to the problem of chemical analysis.

WHAT DO ANALYTICAL CHEMISTS DO?

- Determine the composition or structure of materials.
- Determine how much of a component is present.
- Understand and use the science of sampling, defining, isolating, concentrating, and preserving samples.
- Set error limits.
- Validate and verify results through calibration and standardization.
- Perform the science of separations based on differential chemical properties.
- Create new ways to make measurements.
- Carry out the process of making measurements.
- Interpret data in proper context.
- Communicate results clearly.

SKILLS AND CHARACTERISTICS

Analytical chemists need skills and characteristics that enable them to do the following:

3052–8/95/0067$08.00/0 ©1995 American Chemical Society

- Operate instruments and make repairs to them.
- Interpret and apply technical and vendors' literature.
- Design and carry out measurement-related experiments; define complex problems and devise new ways to measure them using quantitative understanding of measurement processes, statistics, and probabilities.
- Obtain, develop, and interpret data.
- Make decisions and recommendations based on data analysis.
- Manage several projects at once.
- Routinely communicate results in writing and orally.
- Interact regularly on a technical basis with chemists in other specialties.
- On occasion, represent organizational needs to vendors.
- Be oriented toward the customer and provide service to others.

Because analytical chemists often perform a service-based job, expectations involving job function and performance are also based on customer and server viewpoints. We list here the skills and tasks expected of an analytical chemist in a measurement service laboratory. To be successful working with many people and expectations, the analytical chemist must be well-balanced and mature. The skills and tasks from a service perspective include the following:

- Provide a sample analysis service.
- Review a submitted sample for analysis, then run it, or make appropriate referrals if needed. Know which measurements will solve the problem and who makes them. Participate in cross-functional work teams.
- Obtain, develop, and interpret data correctly.
- Communicate results, decisions, and recommendations directly to the customer in readily understandable terms.

KNOWLEDGE REQUIRED

Analytical chemists are expected to have a scientific understanding of sampling procedures; the validity of measurements, calibration, and standardization; and the need for accurate record-keeping. In addition, analytical chemists with advanced degrees are expected to possess and be able to apply one or more of the specific bodies of knowledge listed below. (A Ph.D. is attainable in many of the individual areas listed; each technique is normally associated with specific theories, instrumentation, and protocols.)

Molecular Spectroscopy

- UV–visible
- Infrared
- Raman
- Microwave
- Nuclear magnetic resonance (NMR)
- Electron spin resonance (ESR)
- Mass spectroscopy
- X-ray powder diffraction
- X-ray crystallography

Atomic Spectroscopy

- Atomic adsorption
- X-ray diffraction
- Inductively coupled plasma (ICP) mass spectroscopy

Separations

- Gas chromatography
- Liquid chromatography
- Ion chromatography
- Electrophoresis
- Supercritical fluid

Mathematics

- Statistics
- Chemometrics
- Computer simulation

- Computer modeling
- Probability

Morphology and Surface Science

- Electron spectroscopy for chemical analysis (ESCA)
- Secondary-ion mass spectrometry (SIMS)
- Auger electron spectroscopy (AES)
- Electron energy loss spectroscopy (EELS)
- Brunauer–Emmett–Teller (BET) equation for surface area
- Transmission electron microscopy

Polymer Characterization

- Size exclusion chromatography
- Light-scattering microscopy
- Osmometry
- Field-flow fractionation

Elemental Analysis

- Microanalysis
- Wet chemical
- Radiochemistry
- ICP mass spectrometry
- Emission spectrometry

Instrumentation

- Electronics
- Physics
- Robotics
- Computers (specific applications; general use of computers is assumed)
- Marketing and sales support

Others

- Physical testing
- Electrochemistry
- Photoacoustic and laser spectroscopies
- Chemometrics
- Thermal analysis (differential thermal analysis, differential scanning calorimetry, and thermal mechanical analysis)
- Hazards analysis

WHERE DO ANALYTICAL CHEMISTS WORK?

- In measurement service laboratories—state, federal, and private.
- As generalists or specialists within R&D laboratories of chemical companies and state and federal governments.
- In instrument companies designing and selling instruments.
- In academic teaching and research.
- Because the need for measurement is pervasive in chemistry, jobs for analytical chemists are found in almost all of the areas listed in the introduction to this book.

General Observations

- Whether one has a bachelor's or master's degree seems to have little influence on the pay maximum. The M.S. chemist may have a higher starting salary but may eventually earn no more than a B.S. chemist.
- Forensic positions are often on-the-job training positions.
- Analytical chemists have rather good career security.
- Salaries are similar to those for workers with bachelor's degrees in industry laboratory positions.
- There is some job specialization in labs according to the type of testing done, but no pay differential for one type of testing vs. another.
- Workplaces include the federal government—Federal Bureau of Investigation; Drug Enforcement Agency; Alcohol, Tobacco and Firearms—state crime labs; and independent labs.

AREAS FOR TRANSITION

Laboratory Jobs

- Physical chemistry
- Clinical chemistry
- Environmental chemistry
- Forensics
- Instrumentation development
- Weapons and defense
- Marine science

Nonlaboratory Jobs

- Technical communications areas such as information science and patent support.
- Supervisor of testing and control laboratories.
- Interviewers (for prospective employees) or human resources areas.
- Writer and critic (use for such talents is found in government laboratories, e.g., Environmental Protection Agency and National Institutes of Health, and government service laboratories, e.g., Oak Ridge National Laboratory, Naval Research Laboratory, and Argonne, where many significant procedures must be written and tested).
- Customer services (sometimes a laboratory position, sometimes not) and marketing and sales of analytical equipment.

ℰ℺ Cℬ
CHAPTER 10

PHYSICAL CHEMISTRY

Physical chemistry is the science of probing the nature of bonding between atoms, and modeling and measuring the rates of chemical reactions. Physical chemistry links physical principles with chemistry.

WHAT DO PHYSICAL CHEMISTS DO?

- Play important roles in *all* disciplines of chemistry.
- Study the properties and effects of intermolecular interactions.
- Work in biochemistry, materials chemistry, electrochemistry, kinetics, quantum mechanics, thermochemistry, surface chemistry, polymers, spectroscopy, solid-state chemistry, and theoretical chemistry.
- Measure energy exchanges that occur in chemical reactions.
- Unravel the relationships between electronic structure and molecular properties through experimentation and theory.
- Predict structural and energetic information on molecules based on quantum mechanical calculations.
- Measure and predict likely products, rates, and mechanisms of chemical reactions based on kinetic and thermodynamic data.
- Predict reaction pathways that can be used to control reactions to maximize wanted products or suppress unwanted products.
- Study the properties of binary and multicomponent mixtures and phases for separations.

3052–8/95/0073$08.00/0 ©1995 American Chemical Society

- Study and characterize the rates of catalytic reactions and develop new catalysts.

SKILLS AND CHARACTERISTICS

- High-level problem-solving and analytical skills.
- Computer interfacing and computer programming (data and information analysis, modeling, and simulation calculations).
- Instrumentation—use and development for chemical analyses and characterization.
- Creativity—developing new approaches to problems; for example, making a particular process more cost-efficient.
- High-level math skills.
- Mechanical skill—working with one's hands.
- Data analysis and interpretation.
- Customer service (often serving as resource person, giving guidance to other chemists on data analysis, etc.; timeliness of response is important).

KNOWLEDGE REQUIRED

- Math through calculus and differential equations essential.
- Knowledge of programming for computer use essential (e.g., FORTRAN).
- Some knowledge of electronics helpful, such as knowing how instruments collect data, how signals are processed, and how information is digitized.
- Strong background in theoretical chemistry and physics very useful.
- A broad overview of chemistry and its applications.

WHERE DO PHYSICAL CHEMISTS WORK?

- In chemical companies working in such areas as catalysis, surfactants and colloids, biophysics, environmental chemistry, electrochemistry, polymers, computational chemistry, radiation chemistry, analytical chemistry (including spectroscopies), high-temperature chemistry, materials science, and combustion chemistry.

- In instrument companies, designing instruments.
- In computer companies, specializing in software, such as software design.
- Academic teaching and research.
- In government labs doing R&D work in such areas as laser science, surface science, molecular physics, radiation physics, thermophysics, molecular spectroscopy, environmental chemistry, combustion chemistry, high-temperature chemistry, materials science, nuclear chemistry, biophysics and biotechnology, solution chemistry, geochemistry, semiconductor chemistry, superconductor chemistry and atmospheric chemistry.

Typical Tasks

- Using high-technology instrumentation such as lasers, supercomputers, molecular beam systems, ultrahigh vacuum apparatus, and high-pressure, high-temperature cells.
- Developing advanced instrumentation.
- Analyzing data to determine quantitative understanding of chemical systems, including knowledge of existence and accessibility of relevant information.
- Good knowledge and competent use of personal computers and advanced computational procedures.
- Bridging the gap between fundamental phenomena (e.g., absorption, diffusion, and mass transport) and engineering practice.

Areas for Transition

Opportunities for transitions are readily apparent from the vast array of functions physical chemists perform. In many instances it has been reported that physical chemists are known as some other type of chemist because of their integration into other fields.

&) C&

CHAPTER 11

INORGANIC CHEMISTRY

Inorganic chemistry is the study of the synthesis and behavior of chemical systems that can contain elements from the entire periodic table. It has application in every aspect of the chemical industry, especially in catalysis, materials science, medicine, agriculture, and the environment.

WHAT DO INORGANIC CHEMISTS DO?

- Create and study inorganic and organometallic compounds.
- Create and study chemical reactions that produce inorganic and organometallic compounds and materials.
- Study kinetics and mechanisms of reactions involving inorganic and organometallic compounds and materials.
- Study the chemical and physical properties of inorganic and organometallic compounds and materials.
- Create and study new uses for new or existing inorganic and organometallic compounds and materials.
- Create and implement processes for new or existing inorganic and organometallic compounds and materials.
- Carry out chemical, physical, and solid-state characterization of inorganic and organometallic compounds and materials at the molecular, aggregate, and assembled levels.

3052–8/95/0077$08.00/0 ©1995 American Chemical Society

- Test and model chemical and physical properties of inorganic and organometallic compounds and materials and their interactions with their surroundings.

SKILLS AND CHARACTERISTICS

- A good working knowledge of the chemistry of all the elements.
- A good working knowledge of organic chemistry.
- A good working knowledge of preparation, isolation, and analysis of inorganic compounds, including good experimental skills.
- Working knowledge of literature of inorganic or organometallic chemistry.
- Computer literacy, knowledge of computer modeling.
- Knowledge of nomenclature.
- Ability to work as part of a team, collaborating with colleagues in other disciplines.

KNOWLEDGE REQUIRED

- The chemistry and biological activity of inorganic and organometallic compounds and their effect on plant growth.
- Analytical techniques for the quantitative detection of inorganic and organometallic chemicals (and their breakdown products) used in soil treatment.
- Governmental regulations governing inorganic and organometallic compounds used to treat soil.
- Formulation and application techniques for inorganic and organometallic compounds used in treating soil.
- Industrial hygiene, particularly as related to safe handling and application of inorganic and organometallic compounds used in treating soil.

WHERE DO INORGANIC CHEMISTS WORK?

- In large chemical companies working on catalysts, pigments, surfactants, coatings, pharmaceuticals, plastic materials, and additives.
- In manufacturing plants doing process development and scale-up, especially in catalysis.

- Academic teaching and research.
- In government laboratories doing R&D in many areas, for example, rocket propellants, explosives, pollution control, and catalysis.

AREAS FOR TRANSITION

Laboratory Jobs

- Fertilizer and soil chemistry is the chemistry of inorganic compounds. These chemists are employed by industry and government for research on and the sale and use of inorganic chemicals for treating soil.
- Environmental chemistry
- Fuel chemistry
- Geochemistry
- Medicinal chemistry
- Petroleum chemistry
- Polymer chemistry

Nonlaboratory Jobs

- Patent law
- Regulatory law
- Sales
- Solid-state characterization

&ᔥᔥ

CHAPTER 12

ORGANIC CHEMISTRY

With few exceptions, organic chemistry is the science of synthesizing, characterizing, and applying molecules that contain carbon–carbon bonds; applications exist in a vast array of areas including organic, organometallic, and bioorganic molecules.

WHAT DO ORGANIC CHEMISTS DO?

- Create and study organic and bioorganic compounds, the chemical reactions that produce them, and their chemical and physical properties.
- Test and model chemical and physical properties of organic and bioorganic compounds and their interactions with their surroundings.
- Create and explore new uses for new or existing organic and bioorganic compounds and solutions.
- Carry out chemical, physical, and solid-state characterization of organic and bioorganic compounds and solutions at the molecular, aggregate, and assembled levels.

SKILLS AND CHARACTERISTICS

- Design synthesis reactions (on paper or with computers) using known chemistry.
- Design synthesis reactions using novel chemistry. Create new chemical reactions.
- Model and test synthesis reactions.

3052–8/95/0081$08.00/0 ©1995 American Chemical Society

- Design and test scale-up procedures and processes.
- Carry out synthesis reactions in a laboratory environment using unique and sophisticated chemicals, hardware (often glassware), and associated controlling electronics.
- Carry out product isolation in a laboratory environment using unique and sophisticated chemicals, hardware (often glassware), and associated controlling electronics.

KNOWLEDGE REQUIRED

- Ability to synthesize organic and bioorganic compounds.
- Ability to perform tests using sophisticated instrumentation and interpret results.
- Basic knowledge of polymer chemistry.
- Computer literacy and knowledge of computer modeling.
- Knowledge of nomenclature.
- Ability to work as part of a team, collaborating with colleagues in other disciplines.

WHERE DO ORGANIC CHEMISTS WORK?

- In large chemical companies working on pharmaceuticals, consumer products, surfactants, paints, pigments, and coatings.
- In manufacturing plants doing process development and scale-up.
- Academic teaching and research.
- R&D in small start-up biotech firms.
- In government laboratories doing R&D in many areas, for example, drug synthesis, agrochemicals, and explosives.
- In instrument sales.

AREAS FOR TRANSITION

- Polymer chemistry
- Environmental chemistry
- Silicone chemistry

In addition to the general skills listed for organic chemistry, the following specialties require additional skills for career transitions, many of which may be acquired on the job.

Medicinal Chemistry

- Knowledge of the chemistry and biological activity of organic compounds that make them useful in medicine.
- Research, discovery, and development of new medicinal agents.
- Knowledge of enzymology, biochemistry, genetic engineering, pharmacokinetics, computer molecular modeling, and toxicology.
- Ability to perform assays.
- Ability to analyze clinical trials.
- Knowledge of FDA regulations.
- Ability to perform and interpret mass spectrometry, NMR, HPLC, GC, IR, and X-ray crystallography.
- Knowledge of formulation.

Carbohydrate Chemistry

- Application of organic chemistry techniques to carbohydrates, for example, modifying starches or sugars.
- Knowledge of wet chemistry.
- Experience with any or all of these analytical techniques: mass spectrometry, NMR, IR, GC, HPLC.
- Experience using *Methods in Carbohydrate Chemistry* as a reference.

Agrochemistry

- Knowledge of plant biology and physiology.
- Knowledge of basic toxicology, agronomy, entomology, and soil chemistry.
- Knowledge of EPA regulations regarding pesticides.
- Awareness of the interface between agricultural chemicals and the environment, for example, leachability, degradation, and composting.

- Knowledge of pesticide metabolism.
- Computer molecular modeling.

Agricultural and Food Chemistry

- Knowledge of food science, microbiology, polymer chemistry, carbohydrate chemistry, chemical engineering, plant physiology, nutrition, rheology, and flavor chemistry.
- Knowledge of biological processes and metabolic pathways.

Fuel Chemistry

- Knowledge of the chemistry of fossil fuels (coal, oil, biomass).
- Knowledge of biochemistry or microbiology for biomass or pollution control.
- Knowledge of inorganic chemistry and catalysis for coal liquefaction; co-utilization of fossil fuel and wastes or byproducts; petroleum refining; or pollution control.
- Computer modeling for fuel conversion and petroleum refining.
- Grant proposal writing (much of the work is sponsored by the U.S. Department of Energy, and grants are a major source of funding).

Petroleum Chemistry

- Multidisciplinary knowledge embracing organic, organometallic, inorganic, analytical, polymer, and physical chemistry.
- Basic knowledge of chemical engineering and mechanical engineering.
- Knowledge of catalysis for preparation, characterization, regeneration, reactivation, and reworking.
- Basic knowledge of economics.

Rubber Chemistry

- Knowledge of synthesis, characterization, and applications of elastomeric polymers.
- Knowledge of free radical chemistry.
- Basic knowledge of chemical engineering and mechanical engineering.

- Basic knowledge of ecology.
- Basic knowledge of economics, market forces, and the rubber industry as a business.

Fluorine Chemistry

- Knowledge of safety rules and regulations regarding the handling of fluorine, which is very reactive, is essential.
- Experience in fluorine chemistry is more likely to be gained in an industrial setting in the United States. (There are many more academic fluorine chemistry programs overseas.)
- Knowledge of which principles of organic chemistry can and cannot be applied to fluorine.

ꝏ ლ
CHAPTER 13

BIOCHEMISTRY AND BIOTECHNOLOGY

B iochemistry is the chemistry of proteins, enzymes, carbohydrates, fats, and nucleic acids and their roles in the structure and func- tion of living organisms, systems, and processes. Biotechnology is the commercial application of biological organisms, systems, or pro- cesses by various industries.

Biochemists and biotechnologists generally fall into two work groups: academe and government labs, and industry. In academe and government labs, chemists are primarily interested in long-term, high-risk work with no immediate application. Both basic and ap- plied research are undertaken. Examples of basic research include determination of nucleic acids in genes or purification and charac- terization of an enzyme. Examples of applied research include insert- ing genes into an organism to make multiple copies, producing copious amounts of a gene product (i.e., a protein), and immobilizing an enzyme for continuous use and reuse. In industry, the research is shorter term, and the development is motivated by a desire for profit.

WHAT DO BIOCHEMISTS AND BIOTECHNOLOGISTS DO?

- Develop unique bioprocesses for production of drugs, foods, and chemicals.

3052–8/95/0087$08.00/0 ©1995 American Chemical Society

- Devise new methods for the creation, separation, and purification of biomolecules.
- Determine the structure of biomolecules using interdisciplinary techniques.
- Develop novel methods to determine the structure–function relationship of biomolecules.
- Study the molecular mechanisms by which biomolecules perform their native functions.
- Develop new methods for the analysis of drugs, hormones, and other important biologically active compounds.
- Devise methods to discover new drugs with therapeutic potential.
- Develop novel uses for existing biological compounds.
- Use recombinant DNA technologies to genetically engineer biomolecules with novel properties.
- Design and construct chemical sensors and biosensors for the rapid, sensitive, and precise detection of toxic compounds in the environment, secondary metabolites in the body, and gases in the air.
- Study and modify molecular properties, such as stability, enzyme activity, binding affinity (e.g., hormone, effector, or drug), and physiology effects (e.g., inhibition of viral replication).
- Characterize and optimize bioprocesses for commercial scale-up.
- Validate the manufacturing process by showing that the process consistently delivers a product of the desired quality.
- Assist regulatory affairs professionals in obtaining product marketing approval from government agencies.

SKILLS AND CHARACTERISTICS

- Skill in working with small quantities of materials.
- Creativity and connectivity skills (using intuitive skill to find connections).
- Good observational skills.
- Computer use (databases, searching, and literature work are very important because the field is evolving so rapidly).
- Driven by a sense of mission.

KNOWLEDGE REQUIRED

B.S., M.S., and Ph.D. degrees in a scientific or engineering discipline are desirable plus a good grounding in basic sciences such as chemistry, mathematics, biology, and physics. Postdoctoral work is generally considered valuable. In addition:

- Basic understanding of biochemistry.
- Basic understanding of microbiology, metabolism, and physiology of organisms.
- Basic understanding of nucleic acids and proteins.
- Proficiency in genetic engineering.
- Proficiency in sterile techniques (process validation would require more in-depth knowledge).
- Understanding of radiotracing techniques.
- Appreciation of three-dimensional structures.
- Experimental design.
- Use of intuitive knowledge important in this field.

INFORMATION ABOUT SPECIALTY

There is a critical crossover of skills between analytical chemistry and biochemistry. A large fraction of what is done in biotechnology and molecular biology is really analysis, although it is often done by different techniques, such as electrophoresis and column chromatography. The peptide and protein characterization fields are populated by chemists who are in effect analytical biochemists.

In small biotech firms, the work environment is often academically oriented in the research area, but nonetheless it is a result-oriented culture that places a high value on work hours and commitment to work. Chemists in this work community are mobile and generally willing to move to advance.

Being able to work effectively as a team member is especially critical when doing drug discovery.

AREAS FOR TRANSITIONS

- Agriculture
- Energy

- Environment
- Medicine
- Engineering
- Carbohydrates
- Teaching

ଚ୍ଚ ଓଃ
CHAPTER 14

POLYMER CHEMISTRY

W ith some exceptions, polymer chemistry[1] is the science of de-signing, developing, studying, and characterizing organic mac-romolecules, their aggregates, assemblies, morphology, and applications. It is influenced by the specific subset of chemical theory, chemical reactions, applications, and instrumentation that is consistent with these macromolecules, their design, synthesis, processing, fabrica-tion, and utilization.

WHAT DO POLYMER CHEMISTS DO?

- Design chemical reactions that produce new monomers and polymers.
- Design and synthesize molecules, which comprise monomers and polymers.
- Study the chemical and physical properties of polymer molecules, ag-gregates, and assemblies.
- Make modifications to polymer assemblies (e.g., plasticize, vulcanize, and crystallize).

---

[1]Polymer chemists often refer to themselves as polymer scientists and to their field as polymer science. We use the terms polymer science and polymer chemis-try interchangeably here; however, from the authors' point of view, the term polymer science is broader and includes more of the analytical, applied, and en-gineering sciences in its operations than does the term polymer chemistry.

3052–8/95/0091$08.00/0 ©1995 American Chemical Society

- Formulate polymers for specific applications by combining polymers and additives.
- Develop new uses for existing polymers and properties.
- Develop new processes for assembling polymers into useful products.
- Characterize macromolecules at the molecular and aggregate and assembled levels.
- Convert polymers to end-use products, including end-use evaluation.
- Model and test chemical and physical properties of polymers and their interactions with their surroundings.

SKILLS AND CHARACTERISTICS

- Perform organic chemistry.
- Perform physical and analytical chemistry.
- Perform an engineering role.
- Exhibit knowledge of catalysts.

KNOWLEDGE REQUIRED

- B.S. chemists' knowledge of polymer science varies widely with the degree obtained, and they are expected to learn needed skills on the job.
- M.S. chemists are expected to know the basic concepts of polymer science and to be able to contribute in the laboratory or plant.
- Ph.D.s are expected to have working familiarity with the concepts of polymer chemistry, to be able to initiate new scientific studies or create new products, and to be able to lead others. Ph.D. polymer chemists are expected to have familiarity with all of the following areas and to be state-of-the-art with some:
  - The structure and properties of polymers; analysis and testing of polymers for solubility, molecular weight, molecular size, morphology, crystallinity, rheology, and mechanical properties.
  - The interactive relationships between process, structure, and properties.
  - Polymerization reactions, condensation reactions, free radical and non-radical chain (addition) polymerizations, group transfer poly-

merization, copolymerization, homogeneous systems, heterogeneous systems, and others.

- Polymer processing.
- Plastics technology, fiber technology, and elastomer technology.
- Properties of commercial polymers. Olefins, amides, vinyls, cellulosics, and fluoropolymers.

WHERE DO POLYMER SCIENTISTS WORK?

- Within R&D laboratories of chemical companies and state and federal governments. Some jobs are located wholly in laboratories, others at the desk, at the library, or in computer centers.
- Academic teaching and research.
- Physical test laboratories.
- Semiworks or plants.
- Laboratory, semiworks, or plant in customer facilities.
- Marketing and sales.

AREAS FOR TRANSITION

- Organic chemistry, with mechanism studies and synthesis.
- Physical and analytical chemistries in the areas of physical properties and measurement.
- Engineering in the areas of manufacture and processing.
- Patent support functions, especially by large companies with polymer product lines and by universities and institutes specializing in polymers.
- Computer modeling and engineering design.
- Customer service, in problem-solving and in the marketing and sales of polymer products.
- Silicone chemistry.

ⅇↄ ⒞ℨ
CHAPTER 15

ECOLOGICAL CHEMISTRY

The fields of chemical health and safety, environmental chemistry, and geochemistry are interrelated and seek to protect people and the Earth they inhabit. We have chosen to group these fields together under ecological chemistry.

CHEMICAL HEALTH AND SAFETY

Specialists in chemical health and safety are concerned with the health of other scientists, with the safe use and handling of chemicals, and with the safe operation of instruments and machinery.

WHAT DO CHEMISTS IN CHEMICAL HEALTH AND SAFETY DO?

- Work with industrial hygienists to ensure that employees work safely and that safety and health rules are obeyed.
- Ensure compliance with OSHA laws.
- Ensure proper labeling for chemicals being used.
- Develop strategies for waste management and disposal.
- Monitor the storage of materials to prevent accidents and spills.
- Examine control methods being used.
- Conduct air monitoring studies in the workplace.
- Train workers to understand the importance of health and safety.

3052–8/95/0095$08.00/0 ©1995 American Chemical Society

- Perform safety audits and inspections; participate in safety and health reviews before running new processes or handling chemicals.
- Work with attorneys in liability cases involving health and safety issues.
- Conduct indoor air quality studies.
- Work with forensic issues related to the workplace.

SKILLS AND CHARACTERISTICS

- Persuasion and negotiation skills, and the ability to work with a variety of people from factory workers to management.
- Diplomacy and tact.
- Attention to detail.
- Organizational skills to deal with mountains of paperwork associated with federal and state regulations.
- Knowledge of psychology of teaching and of risk description.
- Ability to build relationships with regulators so that advice can be readily obtained from EPA and OSHA.
- Clearly developed ethical boundaries, i.e., know for whom you are working. In many cases, an in-house safety officer knows of things he or she cannot speak about to other employees.
- Knowledge of risk communication, especially to nonchemists.

KNOWLEDGE REQUIRED

- Good knowledge of OSHA and EPA regulations, including the ability to interpret regulations and work closely with legal staff on matters involving safety.
- Broad knowledge of the periodic table and the ability to predict hazardous properties of elements and their compounds.
- Knowledge of ergonomics.
- Persuasion and negotiation skills.
- Knowledge of OSHA and EPA state regulations and specific federal regulations such as EPA Hazardous Waste Operations and FIFRA.

DESIRABLE CERTIFICATION

Certified Industrial Hygienist (CIH)

CIHs may be toxicologists, chemists, medical doctors, or occupational health nurses. Certification requires written examination based on knowledge acquired in the field plus five years of practical experience. Most chemical health and safety specialists have B.S. degrees, although there is a trend toward making the M.S. degree an entry requirement.

Occupational Safety and Health Technologist

Certification as an occupational safety and health technologist takes three years; 35% of work time must be devoted to safety functions. In small plants and laboratories, the safety and health technologist may be responsible for safety, with oversight provided by corporate headquarters. Typical tasks include monitoring to ensure exposures are within regulations or corporate requirements, and maintaining records.

WHERE DO HEALTH AND SAFETY CHEMISTS WORK?

- 40% in manufacturing or industrial plants
- 40% in the insurance industry
- 10% in consulting
- 10% in other places such as government, state OSHA plans, hospitals, railroads, airlines, and utilities

REPRESENTATIVE JOB TITLES IN THE CHEMICAL HEALTH AND SAFETY SPECIALTY

- Safety officer (government)
- Safety engineer, safety manager, director of safety and health (industry)
- Loss control engineer, loss prevention engineer (insurance)
- Safety and health technologist (typical title for a chemical technician in occupational health and safety)

ENVIRONMENTAL CHEMISTRY

Environmental chemists investigate the source, fate, control, and effect of chemicals in natural and engineered environments. Environmental chemists apply knowledge and tools from the major areas of chemistry (organic, physical, inorganic, and analytical chemistry) and integrate them with other disciplines and subdisciplines such as geochemistry, limnology, oceanography, toxicology, health and safety, and environmental engineering.

WHAT DO ENVIRONMENTAL CHEMISTS DO?

- Study the thermodynamic and kinetic characteristics of processes that affect chemicals in the environment and in engineered treatment systems.
- Study how chemical, physical, and biological processes interact.
- Determine the sources and fates of chemicals in the environment.
- Develop and use analytical techniques to determine concentrations, speciation, and areal and temporal distribution of chemicals in environmental systems.
- Help in risk assessments of toxicants, including mechanism of action and exposure amount, timing, and bioavailability.
- Elucidate historical chemical inputs through the study of chemicals indicative of these inputs.
- Contribute to new or improved industrial processes and treatment systems to minimize undesirable effluents.
- Investigate chemical characteristics of the atmosphere, hydrosphere, and lithosphere, including natural and human-induced variations.
- Collect, measure, predict, and categorize physical, chemical, and bioactivity values for chemicals of interest.
- Model the fate and effect of chemicals in environmental systems.
- Assess financial risk. For example, an environmental chemist working with underground storage tanks or real estate transactions will be involved with the financial aspect of all situations he or she investigates.
- Manage hazardous waste, including control of pollution sources; nonhazardous waste management; environmental management systems operations, performance evaluations, life cycle assessments, and auditing; and recycling and reuse.

- Identify and categorize of waste and hazard.
- Develop systems to separate or isolate waste and to dispose of waste legally.
- Manage air quality, including indoor air quality management, atmospheric chemistry, origin and remediation of air-borne toxins, and air pollution control.
- Manage water quality: monitoring drinking water quality and investigating interactions of drinking water parameters.
- Determine toxic pollutants in surface water and groundwater; work with pollution control and remediation.
- Manage soil and seashore pollution (see also geochemistry).
- Analyze residues.
- Isolate, identify, and quantify breakdown products, particularly agrochemicals and food chemicals.

Skills and Characteristics

- Instrumentation skills.
- Attention to detail.
- Computer use.
- Customer service orientation, especially in laboratories serving a wide range of clients.
- Negotiation skills—important if working with regulatory or legal personnel.
- Ability to work with confidential information.

Knowledge Required

- In the laboratory, B.S. and M.S. chemists primarily do analytical work; generally Ph.D.s manage laboratories, supervise B.S. and M.S. chemists, do troubleshooting and repair of instruments, and interact with regulatory people.
- Some courses in chemical engineering are helpful; at least a knowledge of chemical engineering principles and vocabulary is very helpful. Some toxicology background and an understanding of risk assessment

principles are also useful. In regulatory work, law courses including international law are desirable.

- A degree or experience solely in chemistry may make a transition to environmental chemistry difficult; a year of university training in some aspect of environmental chemistry will enhance possibilities.

- A background in good management practices (GMPs) as related to good laboratory practices (GLPs) may be useful if you are moving from a production position.

Additionally, knowledge of the following environmental laws would be helpful:

- Clean Air Act and 1990 amendments (CAA)
- Resource Conservation and Recovery Act of 1976 and amendments of 1984 (RCRA)
- Comprehensive Environmental Response, Compensation and Liability Act (CERCLA)
- Toxic Substances Control Act of 1976 (TSCA)
- Superfund Amendments and Reauthorization Act (SARA)
- Occupational Safety and Health Act (OSHA)
- Federal Insecticide, Fungicide and Rodenticide Act (FIFRA)
- Emergency Planning and Community Right-to-Know Act (EPCRA)
- Transportation regulations
- Clean Water Act and amendments (1972 and following)
- Safe Drinking Water Act and its 1986 amendments
- Atomic Energy Act
- Federal Food, Drug and Cosmetics Act

CERTIFICATIONS AVAILABLE IN ENVIRONMENTAL CHEMISTRY

- Asbestos—four to five certifications available in this area.
- Underground storage tank installation.
- Hazardous material management.
- Real estate inspector—investigating liabilities associated with real estate.
- Registered environmental property assessment.

- Certified industrial hygienist.
- Certified hazardous materials manager.
- Certified hazardous waste manager.

TYPICAL TASKS OF ENVIRONMENTAL CHEMISTS

- Planning, which may include research to understand the nature of environmental problems; identification of laws and regulations that apply to the environment; proposal of new or revised regulations to govern pollution and pollution control.
- Regulatory compliance.
- Writing or editing of journal articles and other publications about the environment and pollution control.
- Consulting.
- Research.
- Analytical tests.
- Field sampling.
- Database management.
- Environmental modeling.

INFORMATION ABOUT SPECIALTY

- Identify one or two specialty areas in which to major, i.e., hazardous waste, pesticides, water treatment, groundwater, or risk assessment.
- An M.S. degree in environmental science or a chemistry program that emphasizes environmental science with several engineering courses is helpful.
- If pursuing only a B.S. degree, take courses in microbiology or geology or courses related to the particular area of emphasis.
- An interdisciplinary field with course work in areas such as agronomy, biochemistry, biotechnology, geochemistry, geology, hydrology, industrial chemistry, meteorology, microbiology, soil chemistry, toxicology, waste treatment, business management, computer science, or radio-chemistry.

AREAS FOR TRANSITION

- Chemical health and safety, geochemistry, agrochemicals.
- Geochemistry, fuel chemistry, analytical chemistry, petroleum.
- Chemistry, biotechnology, regulatory law.

GEOCHEMISTRY

Geochemists must have a working knowledge of organic and inorganic chemistry and geology. Many geochemists today are involved in environmental matters because ecology involves geology. They frequently work for petroleum companies, mining companies, governmental laboratories, waste disposal companies, and academia.

KNOWLEDGE REQUIRED

Geochemists should have a thorough knowledge of the following:

- Geology.
- Organic or inorganic chemistry.
- Analytical chemistry techniques.
- Basic physical chemistry.

Additionally, the following knowledge would be useful for specialized situations:

- Biochemistry or microbiology for bioremediation or pollution control.
- Hydrology for waste treatment and disposal or pollution control.
- Polymer chemistry for petroleum geochemistry, enhanced oil recovery, and waste treatment and disposal.

AREAS FOR TRANSITION

- Consulting
- Research
- Analytical testing

଼ ଷ
PART III

NONLABORATORY
CAREERS

❧ ☙
CHAPTER 16

REGULATORY WORK

Regulatory chemists work in the areas of human health and environment, ensuring that health and safety standards are met.

WHAT DO REGULATORY CHEMISTS DO?

- Ensure compliance with laws and regulations that protect human health and the environment.
- Monitor the status of local, state, and federal legislation and regulation.
- Develop written programs, training, and standard operating procedures (SOPs) to comply with laws and regulations.
- Manage risk and liability of the employer.
- Assess and communicate risks.
- Design appropriate prevention and control policies in cooperation with those whose methods are being evaluated for risk ("appropriate" meaning that the policies implemented achieve risk reductions by the least obtrusive yet still effective means and do not unnecessarily impede the objectives of the company under examination).

SKILLS AND CHARACTERISTICS

- Reason through highly complex procedures.
- Perform cost–benefit analysis (upper management).

3052–8/95/0105$08.00/0 ©1995 American Chemical Society

- Manage multiple tasks.
- Converse and understand the jargon of the field.
- Communicate in the field without using jargon.
- Develop training programs for adult learners.
- Deal with stress.
- Respond to unusual situations and schedules.
- Be comfortable with open administrative decision-making; active supervision of regulatory agencies by Congress and the courts; and the threat of budget cuts.
- Have excellent communication skills, written and oral.
- Have excellent computer skills, including familiarity with information exchange through networks and comfort with a variety of hardware platforms, e.g., Macintosh, DOS.

KNOWLEDGE REQUIRED

A chemistry degree is exceedingly useful. A law degree is helpful but not essential. Many organizations specifically seek someone with an engineering degree plus a professional engineer (PE) license. Additionally, regulatory chemists should have a working knowledge of the following:

Laws

- Toxic Substances Control Act of 1976 (TSCA)
- Resource Conservation and Recovery Act of 1976 and amendments of 1984 (RCRA)
- Hazardous Materials Transportation Act (HMTA)
- Federal Insecticide, Fungicide and Rodenticide Act (FIFRA)
- Clean Air Act and 1990 amendments (CAA)
- Clean Water Act and amendments (1972 and following)
- Comprehensive Environmental Response, Compensation and Liability Act (CERCLA)
- Superfund Amendments and Reauthorization Act (SARA)
- Emergency Planning and Community Right-to-Know Act (EPCRA)
- Occupational Safety and Health Act (OSHA)

Agencies

- Nuclear Regulatory Commission (NRC)
- Environmental Protection Agency (EPA)
- Food and Drug Administration (FDA)
- Occupational Safety and Health Administration (OSHA), state and federal
- Department of Agriculture (USDA)
- Department of Transportation (DOT), state and federal
- State, county, and city environmental protection agencies

Subjects

- Hazardous waste generation, storage, treatment, and disposal
- Management of underground storage tanks
- Hazard communication and worker right-to-know
- Asbestos management and abatement
- Lead management and abatement
- Management of mixed waste
- Management of new drug applications (NDA), investigatory new drugs (IND), good laboratory practices (GLPs), and good management practices (GMPs) (within a biomedical or pharmaceutical company)
- Exposure control plans
- Controlled substances
- Hazardous materials handling, storage, and transportation
- Toxicology
- Monitoring of airborne contaminants

WHERE DO REGULATORY CHEMISTS WORK?

Government

Typically, a regulatory chemists works in any agency that conducts research, makes chemical measurements, operates a motor pool, or maintains gardens, including the following:

- FDA's National Center for Toxicological Research

- OSHA's National Institute for Occupational Safety and Health (NIOSH)
- Centers for Disease Control and Prevention (CDC)
- Department of Health and Human Services
- EPA's Health Effects Research Lab
- National Cancer Institute
- National Institutes of Health (NIH)
- Department of Transportation's Research and Special Programs Administration
- Federal Highway Administration
- Department of Defense (DOD), Army, Navy, Air Force
- Fish & Wildlife Service
- Forest Service
- U.S. Department of Agriculture (USDA)
- Department of the Interior, Bureau of Reclamation, National Park Service
- Department of Energy (DOE)

Industry

- Environmental health and safety officer
- Regulatory compliance officer
- Chemical hygiene officer
- Director of environmental management

Academe

- Environmental health and safety officer
- Chemical hygiene officer

AREAS FOR TRANSITION

Areas from which a chemist might make a transition include environmental chemistry, safety and health, chemical information, director of R&D, chemistry instructor, biochemistry, medicinal chemistry, manufacturing, and quality control.

꒞ ꒞

CHAPTER 17

LAW

C hemists with or without law degrees can be found working in legal matters. In this chapter we explore both possibilities.

WHAT DO CHEMISTS WITH LEGAL KNOWLEDGE DO?

Chemists also trained in the law can be attorneys in private practice with large or small law firms, chemical companies, universities, private research foundations, and federal agencies.

Scientists trained in the law can work in the areas described below.

Intellectual Property Matters

- These matters include patents, copyrights, trademarks, and trade secrets.
- A degree in a physical science, such as chemistry, allows an attorney to handle intellectual property matters involving a number of chemical specialty areas. For biotechnology law, training in the specialty is needed.
- Chemists in law firms are frequently called upon to handle intellectual property matters that cannot be handled by in-house corporate attorneys because of the volume of work or the complexity of the case. They also represent small companies and individual inventors.
- In chemical companies, attorneys specialize in that industry's intellectual property interests. Bench chemists can make a transition into law as attorneys or patent agents.

3052–8/95/0109$08.00/0 ©1995 American Chemical Society

Environmental Law

- A degree in physical sciences is desirable. Defending against product or process liability vis-à-vis asbestos contamination, and soil, air, and water pollution are involved in this specialty function. Costs associated with environmental issues are high, and cases often are settled out of court.

- In a federal agency such as EPA legal training in combination with technical expertise is highly desirable. Attorneys must draft the laws and regulations as well as enforce them. Consulting on compliance with these laws is also a necessity in this area.

Food and Drug Matters

- Attorneys working in this specialty frequently are not scientists, although the technical knowledge that chemists or toxicologists bring to this area is valuable to an employer or client. Food and Drug attorneys usually have scientists accompany them to court or to the Food and Drug Administration (FDA) to provide needed technical support.

- In the federal and state sector, FDA attorneys handle aspects of litigation through the discovery process, that is, the preparation of interrogatories and depositions. The prosecution is handled by the U.S. Department of Justice. They also check on compliance with the appropriate laws and regulations.

Scientists who are not attorneys support the intellectual property process in chemical companies by working in the areas described below:

Patent Agent

Patent agents must be familiar with the patenting process. They file and prosecute patent applications. The patent agent's interaction with technical staff is important to describe the invention and argue patentability through written documentation. A patent agent often has charge of the patent portfolio in the company.

Patent Liaison

The patent liaison is usually a research person from the technical community who works with attorneys and R&D; does literature searches;

defines what an invention is; and gathers data to distinguish between new inventions and others in the art. The liaison may also serve a licensing function.

Patent Searcher

The patent searcher is responsible for searching patent literature. In some organizations, this search may be done by a patent agent or by a centralized patent searching unit. For more details, see Chapter 20. The patent searcher uses technical knowledge to support the legal process.

Federal Government

Patent examiners for the U.S. Patent Office carry out patent searches, examine patent applications, and handle cases through appeal.

Private Consultants

Patent translators translate patents. They can be accredited by the American Translators Society. Their clients are patent attorneys and inventors.

Expert witnesses assist attorneys in pretrial events and give testimony during the trial, usually after a complaint has been prepared and served upon the defendant. They assist attorneys by discussing the technical content of a complaint in draft form. They may also assist attorneys by identifying existing and correctable weaknesses of a defendant that would likely be at issue in a future suit.

SKILLS AND CHARACTERISTICS

Attorneys

- Excellent writing skills.
- Analytical ability (which would include logic and reasoning).
- Problem-solving.
- Decision-making.
- Interpersonal skills.
- Thoroughness.

- Attention to detail.
- Good judgment.
- Oral communication.
- Networking is very helpful to bring in new clients to a firm, which is an essential part of an attorney's responsibilities in a law firm or in private practice. On occasion, publishing can bring visibility, and perhaps credibility, to an attorney's capabilities. This is also one means of enlarging an attorney's client base.

Patent Agents

- Interpersonal skills.
- Writing ability.
- Listening skills.
- Analytical ability.
- Questioning skills.
- Connectivity skills (ability to see connections from data provided).

KNOWLEDGE REQUIRED

- For all fields of chemistry, either a B.S., M.S., or Ph.D. is required. A law degree may be obtained generally in three years, full-time, and in four years on a part-time basis. Tuition reimbursement often is available in corporations for chemists wanting to pursue this specialty.
- Intellectual property attorneys and patent agents must take and pass a Patent Bar Exam. Patent agents do not have the right to litigate patent cases, but they do not have to go to law school. An apprenticeship period of five years with a more seasoned patent attorney is not unusual.
- Experience as a patent examiner in intellectual property firms is desirable, and many companies provide not only an opportunity for part-time work in law firms while attending school, but may also help to increase the amount of starting salary.
- Patent agents must pass the same patent bar exam as patent attorneys.
- Patent translators must be fluent in languages and be accredited by the American Translators Society.

- Expert witnesses must be regarded as experts in a field. This reputation can often be achieved by publications in the field. A Ph.D. is often desirable for technical knowledge.

TYPICAL TASKS

Attorneys

These tasks are typical of attorneys regardless of type of practice:

- Prepare patent applications and prosecute them at the Patent and Trademark Office.
- If litigating a case, perform extensive research, prepare interrogatories, take depositions, file motions to dismiss, and appear in court.
- Render opinions about patentability, infringement, and validity.
- Take invention disclosures and discuss the nature of the invention to analyze whether the patent or lawsuit should move forward.
- Market services of the law firm to bring in new clients.
- Work with licensing.
- Write contracts.

Scientists Who Are Not Attorneys

The following tasks are typical for people who work with the intellectual property group, including patents, trademarks, copyright, and trade secrets, but who are not attorneys.

- Prepare patent applications.
- Provide support to attorneys by preparing patent applications and information on prosecution matters, opinions, and oppositions.
- Perform patent enforcement tasks.
- Evaluate inventions.
- Assist with the development of a patent application.
- Review abstracts of all publications in an organization to identify if some aspects should be modified because of patent or trade secret considerations.
- Do patent portfolio valuation by reviewing patents previously issued to determine trends; determine if license is needed; determine whether new patents need to be filed.

- Do patent portfolio development to determine those areas still open that need to be strengthened.
- Develop strategies for where to file, what to file, and how broadly to file.
- Train internal clients on intellectual property matters.
- Search for existing patents and other literature and commercial practice.

AREAS FOR TRANSITION

- Intellectual property work as patent agent or liaison to attorney.
- From environmental bench work into environmental law.
- From biochemical or biotechnological bench work into biotechnological law.
- Any chemistry background would be helpful in intellectual property work as an attorney.
- Environmental law plus technical training in environment and safety health issues would present a good opportunity for consulting.

ജ ൽ
CHAPTER 18

HUMAN RESOURCES

This chapter is an overview of the human resource specialty; some of these functions may be carried out solely at the corporate level; some of the functions may currently not be filled by chemists.

WHAT DO HUMAN RESOURCES SPECIALISTS DO?

- Negotiate, implement, and administer employee benefit programs, including monitoring pertinent legislation.
- Identify and implement strategies for identifying, screening, and referring candidates with regard to existing or anticipated job vacancies within the law. This function also includes consideration of internal transfer of employees to other opportunities within the organization. Orienting new employees may be associated with this function.
- Design and administer training and development programs for all levels of staff.
- Manage the outplacement of individuals whose employment is terminated by the organization.
- Design and administer a job evaluation system.
- Design, revise or acquire, and administer a salary administration system that ensures competitive pay practices.
- In partnership with management, ensure that employee policies are developed and communicated throughout the organization.

3052–8/95/0115$08.00/0 ©1995 American Chemical Society

- In partnership with management, ensure that employee relations are handled consistently, uniformly, and fairly.
- Maintain employee records.
- Ensure that the organization meets all of the necessary federal, state, and municipal regulations regarding equal employment opportunity (EEO) matters, including training and development of staff.
- Participate with management in the strategic planning process.
- Perform organizational development functions, such as work climate surveys and assess and recommend new work structures such as teams.

SKILLS AND CHARACTERISTICS

The skills listed are for those in technical recruiting; additional skills would be needed for other functional areas.

- Computer literacy—applicant tracking systems, on-line management of data, and for responding to employee questions.
- Leadership.
- Ability to handle multiple priorities and work with constant interruptions.
- Communications—oral and written.
- Analytical ability.
- Attention to detail.
- Well-developed networks with management for understanding their needs and employee fits within the various units.
- Awareness of organizational philosophy and ability to communicate that philosophy to job candidates.
- Organizational skills.
- Flexibility and adaptability.
- Problem-solving skills.

KNOWLEDGE REQUIRED

To be successful in human resource management, individuals need to broaden their knowledge in compensation, training and development, benefits administration, and employment and EEO laws. Certification is desirable in such areas as benefits and compensation, for example, CEBS (Certified Employee Benefit Specialist) and the American Compensation Certification Program for Compensation Professionals. Additionally, human resource professionals working in employment should be knowledgeable about the laws listed below. Please note that laws affecting other human resource functions have not been listed, as most chemists seem to enter human resources through employment.

- Americans with Disabilities Act of 1990
- Title VII—Civil Rights Act of 1964
- Immigration Reform and Control Act of 1986
- Drug-Free Workplace Act
- Age Discrimination in Employment Act
- Vietnam Era Veterans Readjustment Assistance Act
- Vocational Rehabilitation Act, Section 503
- At-Will Employment
- Equal Pay Act
- Fair Labor Standards Act
- Applicable municipal and state laws affecting employment

GENERAL OBSERVATIONS

The employment function is where most chemists are reported as working within human resources, although chemists are employed in other functional areas such as benefits, employee relations, and compensation. Additional skills or training are needed to perform effectively in the specialty areas of human resources.

AREAS FOR TRANSITION

Volunteering as a technical person to do technical recruiting can be a very effective way to make a transition into this specialty. Generally, Ph.D. chemists who volunteer as technical recruiters recruit Ph.D.s.

ॐ ෬

CHAPTER 19

MARKETING AND SALES

C hemists in marketing and sales work in companies with a wide variety of products and services.

WHAT DO MARKETING AND SALES SPECIALISTS DO?

Specialists in this field manage and plan the delivery of products and services to meet existing or emerging customer needs and profit objectives of the company. Their work is in the following functional areas:

- *Direct sales and technical services.* The direct link between the organization and its customers to obtain value for goods and services rendered. This is the route by which most chemists enter marketing.

- *Marketing.* Product positioning and product sales management, including pricing, researching, and strategically planning the organization's competitive posture.

- *Market research.* Gathering data about product lines and services, usually in support of the product or brand manager.

- *Market communications.* Using varied media to demonstrate the benefits of the organization's products or capability.

- *New business and commercial development.* Taking new products to the marketplace to meet the goals of the business as rapidly as possible.

- *Customer service.* Order placement, handling of sales for customers (order entry, billing, etc.).

3052–8/95/0119$08.00/0 ©1995 American Chemical Society

- *Distribution.* Sending out products.
- *Telemarketing.* Using the telephone and other forms of communications for other than face-to-face selling.

SKILLS AND CHARACTERISTICS

- Desire to sell and a good knowledge of the product line.
- Willingness to learn new technology as it evolves.
- Self-starter, able to work independently; enthusiastic attitude.
- Often must be willing to travel; sometimes must put sales project ahead of personal interests and family.
- Work well under pressure; trustworthy, good judgment.
- Good communication skills, must have "the gift of gab."
- Must be able to translate knowledge of chemistry to products being marketed.
- Ability to find out who the key people are. Ability to use persuasion, not only to close a sale, but also as a marketer.
- Problem-solving abilities.
- Computer literacy (sales people are expected to carry a computer with them to check inventory of products, refer to file of customers, etc.).
- Creative approaches needed to respond to competition.
- Good personal appearance and grooming.
- Well-rounded personality.

KNOWLEDGE REQUIRED

Chemical knowledge (B.S., M.S., or Ph.D.) is considered helpful and required in most organizations. An MBA or financial management training is considered important. For the marketing functions described above, a full marketing program at a university, with sales or marketing courses, would be considered very desirable.

WHERE DO MARKETING AND SALES SPECIALISTS WORK?

Most chemists enter marketing through sales. Employment opportunities may exist in chemical companies, large and small, as well as in organizations specializing in the sales of chemicals, instruments, services, and so forth. Chemists in sales often work out of their homes or in regional sales offices. Movement into a marketing function generally brings the chemist into headquarters. Compensation may be either base salary plus commission or bonus or straight base salary.

REPRESENTATIVE JOB TITLES

- Sales trainee.
- Sales engineer.
- Technical sales representative.
- Account manager.
- Account executive.
- National sales manager.

ᘓ ᘔ
CHAPTER 20

CHEMICAL INFORMATION

W e have chosen to group the fields of database searching, scientific publishing, and research development under chemical information.

DATABASE SEARCHING

WHAT DO DATABASE SEARCHERS DO?

This area includes research project support. In small industrial firms, chemists may also do patent searching. In academe and special libraries, a masters in library science (MLS) is desirable, accompanied by science course work. A chemist in such a job is considered first a librarian and is paid as a librarian. In a library, you must be a specialist in both printed and computerized information sources. An understanding of the language of chemistry, chemical nomenclature, chemical structures, physical properties, and scientific literature is necessary.

SKILLS AND CHARACTERISTICS

- Computer use: Must be able to access database vendors such as ORBIT, BRS, STN, and Dialog.
- Analytical ability.
- Interpersonal skills.
- Attention to detail.

3052–8/95/0123$08.00/0 ©1995 American Chemical Society

- Writing ability.
- Willingness to continue learning.
- Generalist scientist yet must maintain current knowledge in a large number of disciplines.

REPRESENTATIVE JOB TITLES

- Information scientist (in industry).
- Librarian (in academe and scientific libraries).

SCIENTIFIC PUBLISHING

WHAT DO CHEMISTS IN SCIENTIFIC PUBLISHING DO?

This area includes printed and electronic formats. Opportunities for chemists exist in editing, production, writing, and acquiring manuscripts for publication. The nature of the position depends on organizational requirements. Some prior work in the chemical industry as a bench chemist or a prior position that required science training is required. Some free-lance work may be available.

For acquisition editor positions, a knowledge of contract law is helpful as is the ability to prepare a profit-and-loss statement.

Additional opportunities exist for document analysts, who can contribute to building a database by performing on-line editing of chemical information. Foreign language capability plus expertise in a chemical specialty is considered an asset. This position is a production-oriented position.

SKILLS AND CHARACTERISTICS

General skills and characteristics pertinent to scientific publishing include:

- Good editing and writing skills.
- Attention to detail.
- Interpersonal skills.
- Problem-solving skills.

- Flexibility.
- Computer skills.
- Organizational skills.

Additionally, skills specific to document analysts would include

- Knowing when it is appropriate to seek assistance from others when you do not have specialized knowledge in a subject area.
- Desired characteristics are thinker, planner, cautious, fact finder, logical.

REPRESENTATIVE JOB TITLES

- Acquisitions editor.
- Copy editor.
- Assistant, associate, or managing editor.
- Writer.

RESEARCH AND DEVELOPMENT

WHAT DO CHEMISTS DO IN RESEARCH AND DEVELOPMENT?

In R&D, chemical information specialists use computers as a means to develop ways to provide chemical information useful to scientists. A background in artificial intelligence or expert systems is useful. College work related to research in computers and chemical structures is an asset. Most often, chemists with M.S. and Ph.D. degrees work in the area.

Some of the typical research problems of chemists working in this area include the following:

- Develop methods for storing and retrieving information on specific substances and classes of substances.
- Develop techniques for storing and retrieving three-dimensional representations of structures.
- Find applications for artificial intelligence expert systems and hypertext to chemical information.
- Develop uses for CD-ROM.
- Generate chemical names from structural information.

- Extract data automatically from primary journals.

Skills And Characteristics

The following general skills and characteristics are important to all chemists in R&D:

- Ability to conceptualize computer systems.
- Team-oriented.
- Good oral communication skills.
- Influencing skills.
- Flexibility and adaptability.
- Written communication skills.

Additionally, the following skills and characteristics are specific to software design:

- Knowledge of current software programs to do competitive analysis.
- Knowledge of the core business for which the product is being developed.
- A good sense of visual design to create the user interface.
- Ability to conceptualize.
- Good communication skills, both oral and written.
- Computer skills, especially programming skills.

Some opportunities exist for individuals with strong scientific backgrounds for positions that include a training function. These positions are expected to provide conceptual rather than mechanical training to users.

ॐ ॐ
PART IV

OTHER CAREERS

ဢ ௶

CHAPTER 21

COMPUTER APPLICATIONS IN CHEMISTRY

C omputer applications touch every field of chemistry, and computer literacy is one of the core skills required of chemists. Those chemists working full-time out of a central service unit and on a cost-reimbursable basis require an ability to develop good customer relations. This requirement means that providing high-quality results in a timely fashion is important. There are opportunities for B.S., M.S., and Ph.D. chemists, but the majority of people in this area have a Ph.D.

HOW DO CHEMISTS USE COMPUTERS IN CHEMISTRY?

- Molecular modeling.
- Synthesis design: determining how to make new molecules as well as what molecules to make.
- Predicting properties.
- Determining molecular similarity.
- Quantum chemistry: predicting structures, detailed calculation of physical properties.

3052–8/95/0129$08.00/0 ©1995 American Chemical Society

- Chemical reaction database development.
- Computational biophysics: protein structure prediction.
- Structure determination.
- Structural databases: building, maintaining, and using files such as Brookhaven and Cambridge Databases.
- Experimental design.
- Software design: programming algorithms, data acquisition, and analysis.
- Data management: organizing data generated by individual chemists into a centralized file.
- Chemometrics: a subset of data management, data analysis.
- Desktop publishing to generate reports.
- Editing publications.
- Managing data generated by chemical instruments.
- Product design with respect to instruments.
- Drug design: modeling biological membranes, simulating structural dynamics, permeation of drugs through membranes.
- Process modeling.

SKILLS FOR TASKS INVOLVING DESIGN

- Problem-solving.
- Presentation skills.
- Communication skills, both oral and written.
- Creativity—must be an idea person.
- Must understand industry needs.
- Connectivity skills—must be able to visualize connections between trends and needs of chemists.
- Computer skills.

SKILLS FOR LAB SUPPORT

- Listening skills.
- Customer-service orientation.

- Programming ability in C, FORTRAN, and MATLAB (chemometrics).
- Communication skills, oral and written.
- Presentation skills.
- Computer skills—Windows, Internet, and spreadsheets.
- Math—linear algebra; must not be afraid of numbers.
- Statistics—multivariate statistics.
- Problem-solving.
- Influencing skills to sell ideas about using computational skills to other chemists.

GENERAL OBSERVATIONS

Comprehensive knowledge and competent use of computers and advanced computational procedures are needed in these areas, including programming and use of relevant software.

AREAS FOR TRANSITION

There are some opportunities for full-time consulting in this field, especially managing data generated by computers.

Because computers touch every specialty of chemistry, chemists from every bench specialty have opportunities in this field, provided they have the prerequisite skills.

ജ ശ
CHAPTER 22

CHEMICAL EDUCATION

C hemical educators are employed in secondary schools, community or junior colleges, colleges, and universities. Chemical education encompasses all fields of chemistry; a knowledge of other fields of science and mathematics may be necessary or desirable.

CHEMICAL EDUCATION IN SECONDARY SCHOOLS

Secondary school chemical education is a promising field for chemists. Positions in public schools usually require state certification; requirements for state certification vary from state to state, with some states permitting teaching while certification is being obtained. Teaching in a private or parochial secondary school does not usually require certification.

KNOWLEDGE REQUIRED

- A basic, broad understanding of chemistry with an emphasis on inorganic and organic chemistry.
- A knowledge of the philosophy of education and teaching methodology. The secondary school chemistry teacher must understand the needs and varied learning styles of high school students. For those chemists not familiar with the educational process, motivational technique training is generally included in methodology classes.

- A basic knowledge of another science or mathematics is recommended because many secondary chemistry teachers are expected to teach mathematics or another science.
- The ability to balance many tasks.
- The desire and skill to teach.

ACADEMIC TRAINING

- B.S., M.S., or Ph.D.

CHEMICAL EDUCATION IN COMMUNITY AND JUNIOR COLLEGES

To teach in a community or junior college requires a broad knowledge of chemistry plus a knowledge of other sciences or mathematics. In these schools the emphasis for the employee is on teaching.

KNOWLEDGE REQUIRED

- A broad knowledge of all the major fields of chemistry.
- A basic knowledge of another science or mathematics.
- A knowledge of the philosophy of education and teaching methodology.
- The desire and skill to teach.

ACADEMIC TRAINING

- Usually an M.S. or Ph.D.

CHEMICAL EDUCATION IN SMALL TO MEDIUM-SIZE COLLEGES

To teach in a small to medium-size college requires a broad knowledge of chemistry, specialization in an area of need to the school, and some teaching experience. The emphasis is on teaching, but the teacher is often expected at least to supervise undergraduate research if not to do his or her own research. Publications are not mandatory but very desirable. College teachers are expected to be part of the campus life, department administration, and perhaps even part of the college administration.

Service on faculty committees is mandatory. Teachers are also expected to remain active professionally.

KNOWLEDGE REQUIRED

- Broad knowledge of chemistry with in-depth knowledge of fields of need such as organic, analytical, or physical chemistry.
- Prior teaching experience; prior adjunct teaching positions are very desirable.
- Computer skills, particularly some experience with educational technology and databases.
- Ability to convey to the student the principles of chemistry especially to the nonscience major, i.e., the English or physical education major or prenursing student.
- Patience to be able to instruct novices in chemical research and to be able to do research through inexperienced students.

ACADEMIC TRAINING

- Usually a Ph.D.

CHEMICAL EDUCATION IN LARGE COLLEGES AND UNIVERSITIES

Teaching and conducting research in a large college or university requires a Ph.D. degree or an M.S. degree with experience. Publications are a necessity, and patents are also important. Many schools require applicants to have research proposals in hand. Faculty is expected to participate in departmental administration and to serve on a number of committees.

KNOWLEDGE REQUIRED

- In-depth knowledge of a field of chemistry such as organic, physical, analytical, or polymer.
- In-depth knowledge, experience, and publications in a specialized area of research.

- Ability to direct the research of others; ability to manage large research teams or a large group of independent researchers, including postdoctoral students.
- Ability to do research and publish such research.
- Ability to write research grant proposals; ability to fulfill requirements of such grants when received.
- The desire and ability to teach and to do research, sometimes with inexperienced graduate students.

ACADEMIC TRAINING

- A Ph.D. with postdoctoral training or industrial experience.
- Rarely, an M.S. with vast experience and publications history.
- Political astuteness; ability to recognize changes coming and adjust thrusts accordingly.

೮ೕ ೮ೃ
CHAPTER 23

SELF-EMPLOYMENT

At one time or another, almost all chemists who are unemployed or dissatisfied with their work situation consider going into business for themselves. There are definite advantages to having your own company and being your own boss. These include setting your own time schedules, assuming responsibility, and perhaps most important in the long run, the potential to earn considerably more money than you would as an employee. However, a major fraction of new business ventures fail within a short period of time. This section addresses some of the start-up problems a new business faces and suggests ways to overcome them.

Going into business for yourself seldom means that life will be the same as it has been. The conventional 40-hour week may be gone forever. Normally, if you are going into a retail operation, you must be prepared to operate six or seven days a week. Will the store be open from early in the day until midnight? If it is a manufacturing business, who opens up in the morning, who closes up at the end of the day, and who keeps the books? If it is a service operation, are you prepared to provide service any time of the day or night, weekends, or holidays? Do you have a reliable employee or someone else who can handle the business if you are sick or away from the business on vacation? Above all, are you prepared to face an acute short-

NOTE: The following people participated in preparing this section on small business: Larry Bray, Birchard & Bray, Accountants; Ronald Gray, Chem Service Inc.; Lyle H. Phifer, Chem Service Inc.; and Jay Young, consultant.

age of money because of unexpected events? We discuss here two types of small businesses that chemists frequently enter manufacturing and consulting.

USING THE SMALL BUSINESS ADMINISTRATION

If you are seriously considering starting a small business, be it consulting, manufacturing, or any other, first contact the Small Business Administration. This agency cannot solve all your potential problems, but it does offer constructive advice on information resources, raising capital, and preparing a business plan, and can answer many other questions that you may have. To start, call the Small Business Answer Desk at 1–800–827–5722. Ask for the location and the number of the Small Business Administration office nearest to you. Request a copy of "Focus on the Facts," a series of individual sheets covering areas of interest to prospective owners of small businesses.

MANUFACTURING

Raising Capital

Just as with consulting, if you start a manufacturing company you must have sufficient resources to allow you to pay both business and personal expenses for some time. Because few new businesses break even financially in less than two years, it is crucial to have sufficient income on hand to support yourself and your family. If possible, this money should be from savings rather than from a house mortgage or equity loans. It's fine to be optimistic, but a risk of failure always exists.

Money for business inventory, rent, equipment, and other operating expenses can be obtained from a number of sources. The best sources are personal savings, family, and banks. Remember that any outside investor has a piece of your business, and although it may be necessary to use this source, any other way is preferable. Banks are usually the best source of operating capital. A bank is in business to make money, and the way it makes money is by lending money. But banks are cautious, and they want to know exactly what the money is to be used for and how you intend to repay it.

To get a bank to give you a loan, you must be prepared and organized. You must have a business plan and be a good credit risk. If you are a poor credit risk personally, the odds are against you when dealing with a bank.

Establishing credit with suppliers as well as with a bank is a very important part of any business. Adopt a policy of paying any debt on the day it is due. If you have financial problems, talk to those you owe in person. Do not avoid them or fabricate excuses. Companies rapidly get tired of delay tactics, such as, "We just changed our accounts payable person," "We lost the invoice," or any other of the miscellaneous excuses for not paying them. You will soon lose your credit.

Most financial problems in a small business originate from the cash flow. Unless you are in retail sales, most companies operate on a 30- to 60-day payment schedule. Labor and other overhead costs must be paid promptly, and you must have enough cash on hand to pay these costs while waiting the customary 30–60 days to be paid. If your customers don't pay on time, you may be in trouble. Many customers, including large, well-known companies, as well as state governments, deliberately make you wait 60–90 days. One solution is to demand in advance that payment be COD or guaranteed by a credit card, or just to be very careful to whom you extend credit. In any case, count on the fact that 20–25% of all customers will be late in making their payments, and take this into consideration when calculating your money reserves.

Living with the Government

All businesses are subject to government regulations. These regulations start locally with zoning, local water and environmental regulations, whether you have a septic tank or public sewer, sprinkler systems, waste regulation, and even parking arrangements. On the state level, you are generally concerned with the labor laws and additional environmental regulations. Right-to-know legislation in most states requires supplying Materials Safety Data Sheets (MSDS) for any chemical on your premises. Don't forget state sales and use taxes, which can be very confusing.

At the federal level, any chemical operation has to deal with the Resource Conservation and Recovery Act (RCRA), the Superfund Amendments and Reauthorization Act (SARA), the Occupational Safety and Health Act (OSHA), the Environmental Protection Agency (EPA), and the Department of Transportation (DOT). Unless you are thoroughly familiar with each of these, it is best to hire competent environmental and safety consultants. Violation of federal rules frequently results in substantial fines, even jail. Ignorance of the law is no excuse.

Meeting with an Accountant

An experienced certified professional accountant can be a valuable asset to the small business owner. The nature and extent of the services needed from an accountant vary, based on the particular needs of the business.

Consulting with an accountant before start-up can clarify your options on business structuring and related tax and record-keeping requirements. Well-designed accounting systems must be in place to safeguard the assets of the business enterprise. Additionally, the accounting system must provide timely information that measures operating results and fulfill other reporting requirements.

Business owners must decide the most cost-effective method of implementing and maintaining an accounting system. Typically, this method will involve both internal and external accountants.

Internal accounting functions include all those tasks involved with the daily business operations, such as customer billing and receipts, vendor orders and payments, cash balances and projected requirements, inventory control, and employee record-keeping. In the beginning, it may be most cost-effective to use outside accounting services, with only a minimum of internal accounting. As a company grows, the need for an effective internal accounting system increases.

External accounting services can help you in a variety of ways. They can coordinate with the internal accounting operation to ensure compliance with governmental reporting requirements, while keeping you well-informed of operational results. Using outside accountants to prepare financial statements and tax returns also provides a degree of reliability and independence that is sought by lending and bonding institutions.

The small business owner is faced with many challenges. Accounting is an integral part of the business operation, and it is an area that should enhance ownership's ability to make the timely, knowledgeable decisions necessary for success.

Advertising

Once you have completed your market research and determined who your customers are or will be, you must tailor your advertising to meet the needs of both your marketing and business plans and your customers. Advertising is crucial to the growth of your business in the following ways:

- Attracting new customers.
- Reinforcing your company image with existing customers, leading to repeat and referral business.
- Identifying your products with an industry segment or product type.
- Gaining acceptance and orders from competitive customers.

It is important to realize that even though advertising is responsible for your business growth, it is often the first discretionary expense cut by small businesses when times are tough. The quality of your advertising is a direct reflection on your company and your products. You should never cut back on the quality of your advertising as a way to cut costs.

Advertisements can now be developed on personal computers using good desktop publishing software. More sophisticated advertising, such as four-color ads, should be contracted out to professional advertising agencies or graphic design companies. Although a larger advertising or graphic design company may produce a high-quality advertisement, it may not be the most cost-effective alternative for a small business. Locate a firm or individual who knows your market or product type, is familiar with the trade or industry publications you intend to use, and produces high-quality advertisements at a reasonable cost.

There are alternatives to print advertising for your company and your products. Direct mail or telemarketing can increase the size of your contact database. Trade shows or professional societies can enhance your visibility and name recognition within industry segments. Radio or TV spots have the potential to reach a large target audience.

Finally, just because you do preliminary market research does not mean that the study ends there. Continue to survey both your customers and the market as a whole. You may find that a significant amount of your product is being purchased by customers in a market segment that your initial research considered minor. Technology, regulatory, or use changes may have rendered one of your promising products obsolete while boosting the potential of a product in the development stage. Remember that one of the advantages of being a small business is the ability to implement change on short notice to satisfy customer and industry needs.

POINTS TO REMEMBER FOR THE SELF-EMPLOYED

Diversify.
The importance of establishing a large and diverse customer base is crucial to the success of any new small business. The loss of one customer from a large base then will not capsize the business. Likewise, the damage caused by the loss of a group of customers from a particular segment of your product line can be minimized either through new product introduction or by refocusing existing resources on the remaining product lines.

You Can Fail, Even with a Superior Product.
Entering a market dominated by a well-respected company with reliable products can be a daunting task. Although you may perceive your own products to be technically superior, this may not be important to the customers. Before investing time and money into promoting your product, survey the customers to find out whether they actually need or want what you sell and are willing to pay your price.

Pay Attention to the Codes.
There are a variety of codes, not only the National Electric Code, all of which the entrepreneur must be aware of, and must follow to avoid manufacturing failures or catastrophes.

Labels Are Important.
Know your product and label it accordingly. Failure to do so can lead to in-plant or customer accidents that could cause your company's demise.

Protect Your Business and Watch Your Employees.
Know how to manage your staff to avoid counterproductive activities. Cultivate your employees to encourage loyalty, but make sure they are trustworthy if you plan to entrust them with any proprietary information.

The Law Is the Law.
Do not sidestep regulations that do not seem to make good chemical sense by adopting modifications or shortcuts. The law is mightier than the degree, and it will win out over any chemist's superior knowledge.

Pay Attention. Be Streetwise.
Know your market and be prepared to adapt to changes in it, such as a customer's solvency or plan to diversify. Also, pay attention to what any competitor may be doing.

CONSULTING

Many experienced chemists, at one time or another, have privately contemplated the prospect of an enjoyable and rewarding career as a consultant. A midcareer change offers the opportunity to consider this option more closely. Many books have been written on the subject, some dedicated specifically to the chemical area, such as *Trends in Chemical Consulting* (Sodano and Sturmer, 1991). These sources should be reviewed to obtain more detailed information to consider before making a transition into full-time consulting. The following sections cover a few major topics to assist you in deciding whether the life of a consultant would be a good choice for you.

Requirements

To be successful, consultants must appear to be able to satisfy the needs of prospective clients. Clients usually are seeking help from individuals who possess specific abilities or knowledge that would not otherwise be available to them. Hence, special skills or experience are the most important things that a consultant can offer initially. Generalists are usually at a disadvantage as consultants, compared with chemists with a knowledge of "niche" areas.

Clients also usually want reassurance that their consultants do indeed possess the requisite abilities or knowledge. Therefore, references, whether in the form of technical publications or personal recommendations, are important.

The skills and experience that a consultant offers do not necessarily have to be in current "hot" areas of chemical science. Frequently, clients need assistance in areas that have matured and in which R&D is no longer performed. This need can represent a special opportunity for the older chemist.

Getting Started

It takes a relatively small capital investment to enter consulting. Modern technology has made this possible. Hardware should be limited to essentials, such as a good word processing system and laser printer, business telephone,

---

NOTE: This section was written by Donald J. Berets, consultant and owner of The Chemists Group, a placement firm in Stanford, Connecticut.

answering machine, facsimile machine, personal copier, and reasonable space for a desk and files. Generally, nonessential items, such as rental office space and a secretary, should be avoided, at least initially.

As with computers, software is important to success. A well-prepared résumé and the support of colleagues and organizational affiliations are important. Word processing permits résumés to be customized to suit the perceived interests of prospective clients. Access to appropriate analytical or other lab facilities, although not essential, will greatly expand the scope of consulting cases you can accept.

Marketing

For most chemists, whose names are not already household words in the industry, marketing is an essential element to success as a consultant. Unfortunately, marketing is the area in which most chemists, no matter how expert they are in their special fields, are least prepared and most inadequate. Consultants should expect to spend 25–75% of their time on marketing. Of course, with successful acquisition of clients, more time can be spent in technical consulting and less on marketing.

Networking is generally considered to be the most effective form of marketing for consultants. A network can be composed of colleagues, past and present, as well as former supervisors, subordinates, competitors, customers, other consultants, and academic associates. Having a successful network requires sending résumés and appropriate cover letters to many people, not unlike the effort that must be made to find a permanent job with an employer.

Various groups around the country provide a central marketing function for their consultant members or associates (see the *American Consulting Engineers Council Membership Directory* on p. 158). Consider affiliating with one or more of these; most groups charge minimal or no fees and do not preempt independent consulting outside their auspices.

The most likely initial prospect as a client is your immediate former employer. Efforts to line up a consulting arrangement should be initiated, if possible, while you are still employed. Such an arrangement can sometimes be negotiated as part of a termination package, if the separation is an amicable one. Acquiring an initial client is usually the most difficult step for a new consultant, so starting out with one is a good head start.

An outgoing personality and the ability to withstand rejection are invaluable assets to any consultant. As a result of recent industrial downsizings, significant competition exists for consulting assignments. These days a prospective client often has a wide selection of consultants from whom to choose. Being selected for the assignment is important.

Financial Matters

A consultant's fee must take into account the prior investment in marketing that should be recouped. Therefore, at least for consultations of relatively short duration, typical fees are two or more times the pay rate previously experienced by the consultant as a regular employee. For example, a consultant's former $70,000 per year salary would calculate to $100,000 (including fringe benefits), or about $50 per hour. A consulting fee of $100 per hour is reasonable. Out-of-pocket expenses, if significant, are considered reimbursable. The watchword on fees though is flexibility. If the consultant is very busy, fees can be raised. If the client is economically strapped, a consultant may accommodate with a lower fee, if only as a long-term marketing strategy.

Entering into consulting requires a business plan, even if it is only a rudimentary one. What is minimally needed is an approximation of how many client-hours can be expected in the first few years, an estimate of an average hourly fee rate, an estimate of operating costs, and an estimate of the income that is required to operate a successful consultancy. For some older chemists, with income from savings or pensions, consulting may not have to provide a total income source.

Most prospective consultants are concerned about professional liability and seek some form of insurance. As far as is known, there are no companies that offer liability insurance to chemists on financial terms that are reasonable. Although the risks of being a defendant in a lawsuit as a result of consulting activities are minimal, the best protection against personal financial risk is to incorporate. This is a relatively simple procedure in most states and should be seriously considered.

For most chemists, consulting as a full-time career is difficult. To use a quotation somewhat out of context, "Many are called, but few are chosen." Fewer still achieve a fully satisfactory financial return.

On the positive side, there is considerable professional satisfaction to be derived from consulting. Consultants who are in the right place, with the right background, at the right time can do very well. Perhaps only an experimental test will reveal whether consulting is right for you.

ℬ ℭ

APPENDIXES

∽ ∾
APPENDIX I

RESOURCES

BOOKS

American Association for Retired Persons. *How To Stay Employable: A Guide for the Midlife and Older Worker* (D14945). AARP Fulfillment (EE0434), P.O. Box 22796, Long Beach, CA 90801–5796.

American Association for Retired Persons. *Returning to the Job Market: A Woman's Guide to Employment Planning* (D14952). AARP Fulfillment (EE0434), P.O. Box 22796, Long Beach, CA 90801–5796.

American Association for Retired Persons. *The First Step: Getting Started in Your Own Business* (D15084). AARP Fulfillment (EE0434), P.O. Box 22796, Long Beach, CA 90801–5796.

Beatty, Richard H. 1988. *The Complete Job Search Book.* New York: John Wiley & Sons.

Birch, David. 1987. *Job Creation in America: How the Smallest Companies Put the Most People to Work.* New York: Free Press.

Birsner, E. Patricia. 1991. *Mid-Career Job Hunting: Official Handbook of the 40+ Club.* New York: Simon & Schuster/ARCO.

Block, Peter. 1981. *Flawless Consulting: A Guide to Getting Your Expertise Used.* San Diego, CA: Pfeiffer & Co.

Bolles, Richard Nelson. 1993. *What Color Is Your Parachute?* Berkeley, CA: Ten Speed Press.

Brudney, Juliet F. and Hilda Scott. 1987. *Forced Out: When Veteran Employees Are Driven from Their Careers.* New York: Simon & Schuster.

3052–8/95/0149$08.00/0 ©1995 American Chemical Society

Connor, J. Robert. 1992. *Cracking the Over-50 Job Market.* New York: Penguin/Plume.

Drake, John D. 1991. *The Perfect Interview.* New York: American Management Association.

Gould, Richard. 1986. *Sacked! Why Good People Get Fired and How To Avoid It.* New York: John Wiley & Sons.

Half, Robert. 1990. *How To Get a Better Job in This Crazy World.* New York: Crown.

Irish, Richard K. 1987. *Go Hire Yourself an Employer.* New York: Doubleday.

Jackson, Tom. 1991. *Guerrilla Tactics in the New Job Market.* New York: Bantam.

Kennedy, Jim. 1987. *Getting Behind the Résumé.* Englewood Cliffs, NJ: Prentice Hall.

Kent, George E. 1991. *How To Get Hired Today.* Lincolnwood, IL: VGM Career Horizons.

Krannich, L. Ronald. 1989. *Careering and Recareering for the 1990s.* Manassas, VA: Impact Publications.

Matson, Jack V. 1990. *Effective Expert Witnessing: A Handbook for Technical Professionals.* Chelsea, MI: Lewis Publishers, Inc.

Medley, H. Anthony. 1984. *Sweaty Palms: The Neglected Art of Being Interviewed.* Berkeley, CA: Ten Speed Press.

Molloy, John T. 1988. *New Dress for Success.* New York: Warner Books.

Parker, Yana. 1984. *The Résumé Catalog: 200 Damn Good Examples.* Berkeley, CA: Ten Speed Press.

Petras, Kathryn and Ross Petras. 1993. *The Over-40 Job Guide.* New York: Simon & Schuster/Poseidon.

Pettus, Theodore. 1981. *One on One: Win the Interview, Win the Job.* New York: Random House.

Saltzman, Amy. 1991. *Downshifting: Reinventing Success on a Slower Track.* New York: Harper Collins.

Sodano, Charles and David M. Sturmer, Eds. 1991. *Trends in Chemical Consulting.* Washington, DC: American Chemical Society.

Tepper, Ron. 1993. *The Consultant's Proposal, Fee, and Contract Problem-Solver.* New York: John Wiley & Sons.

Wallach, Ellen J. and Peter Arnold. 1984. *The Job Search Companion.* Boston: The Harvard Common Press.

Wendleton, Kate. 1992. *Through the Brick Wall: How To Job Hunt in a Tight Market.* Vollard Books.

Yate, Martin. 1992. *Cover Letters That Knock 'Em Dead.* Holbrook, MA: Bob Adams.

Yate, Martin. 1991. *Knock 'Em Dead.* Holbrook, MA: Bob Adams.

Yeomans, William N. 1984. *One Thousand Things You Never Learned in Business School: How To Get Ahead of the Pack and Stay There.* New York: McGraw-Hill.

ON-LINE CAREER OPPORTUNITIES

The Internet is simply computer networks that are linked together. There is no central point of administration, and if you have a modem and telecommunications software, or subscribe to an on-line service, you can access the Internet.

As the networks grow, so does the information on them. There are five major parts to the Internet: electronic mail (e-mail), on-line conversations, information retrieval, bulletin boards, games, and gossip. The three parts of most interest to the job seeker are e-mail, information retrieval, and bulletin boards.

E-mail is rapidly expanding as an effective way to communicate with human resources and laboratory managers. Job seekers can send not only a letter of inquiry, but also format a résumé to send over the Internet. Therefore, instead of the résumé and cover letter arriving at a company in three to five days, it arrives in less than two minutes.

Information retrieval is the primary application of the Internet, and the tool most frequently used is Gopher. Gopher organizes information into menus of related information. And, many job postings and job databases are found in Gopher.

If Gopher software is available, it usually has an option that allows the user to access other Gopher servers. If not, it is necessary to "telnet" to a Gopher. One of the better Gophers is located at Michigan State University. To access, type: **telnet gopher.msu.edu**. At the login, type: **gopher** and the type of computer you are using.

Another type of information retrieval tool is the World-Wide Web (WWW). It is similar to Gopher in that it links the user to other networks. But, the WWW allows for better graphical applications, transfers information faster, and still allows links to Gophers. Currently, career-related information is limited on WWW; however, it is quickly expanding. To access WWW, the user can go through the MSU Gopher. Instead of logging in as **gopher**, log in as **web** and the WWW is opened.

Usenet or bulletin boards is the third most frequently used resource on the Internet. Usenet provides discussion groups on a variety of topics ranging from computers to the latest episode of *Northern Exposure*. However, it does have employment-related groups and can also be a valuable tool for networking.

Once in Gopher, it is possible to access all Gophers in the world, and there are thousands of them. To save time browsing through all of them, the University of Southern California has already grouped most of the career Gophers together. Select USC Gopher, and this will bring up another series of menus. The path is as follows:

USC Gopher/
    Other gophers and information resources/
        Gopher jewels/
            Employment opportunities and resume...

The last menu contains a variety of career-related postings and on-line services. While in the WWW, users may still access all the Gopher servers, and in addition, find jobs posted exclusively on the WWW. The best access is through on-line job services, which not only lists WWW postings, but also looks at some Gopher postings. A few examples are:

- Online Career Services (OCC)—information regarding this employment advertisement database can be accessed via Internet at **occ-info@msen.com** or by telephone at (317) 293–6499. You may enter your full-text résumé into the OCC Internet database via e-mail to **occ-resume@msen.com**. The résumé must be in ASCII format. This service is also accessible from most other major on-line networks, e.g., Compu-Serve, Dialog, America Online. OCC is free to applicants. Employers must pay a fee to subscribe and to list job postings.

- Academic Position Network (APN): telnet to Gopher and give the address **staff.tc.umn.edu 11111**.

- *Chronicle of Higher Education's* Academe This Week lists job openings in academe. Root Gopher server: **chronicle.merit.edu**.

- American Astronomical Society Job Register (a WAIS database) **aas_jobs.src**.

- AMI groups together job openings posted on other Gophers.
- Biosci posts from the bionet biology newsgroup.
- *The Scientist* is now available on-line. Users may access it as follows:

  If Gopher software is not available
  Type: **telnet ds.internic.net**
  At login, type: **gopher**
  If terminal type is "unknown," enter a new value or press Return.

  OR
  Type terminal type if you know it, or type: **vt100**
  From successive menus, choose:
  4. InterNIC Directory and Database Services (AT&T)/
  4. InterNIC Directory and Database Services (Public Databases)/
  6. THE SCIENTIST-Newsletter

  If Gopher software is available
  Type: **gopher internic.net 70**
  From successive menus, choose:
  4. InterNIC Directory and Database Services (AT&T)/
  4. InterNIC Directory and Database Services (Public Databases)/
  6. THE SCIENTIST-Newsletter

- Federal Career Opportunities Access (not government owned) database which, for a fee, allows you to download notices of federal job vacancies. Updated weekly. For information regarding this service, call (703) 281–0200.
- Women in Science and Engineering Net encourages women in science, mathematics, or engineering to join, with the objective of improving access to careers and advancement in science for women of diverse backgrounds. Access via **wisenet@uicvm.bitnet**.
- Usenet groups are less organized and will take more browsing to find those that would be useful to specific needs.
- Common worldwide newsgroups are found under **misc.jobs**. This category is broken down into discussions about job hunting, résumé posting, etc. In the United States, it is organized first by **us.jobs** then by state or city, for example, **atl.jobs** (Atlanta) or **fl.jobs** (Florida). Again, depending on the group, the information posted could vary. Before contributing to the group, users should read the Frequently Asked Questions (FAQ) and current postings.

The resources listed above are a starting point; career-related information is constantly being updated. Therefore, users should not limit themselves with the above information. For users who are not very familiar with the various Internet resources, or have only just begun to explore the Internet, the following reference materials will help:

- Levine, John R. and Carol Baroudi. 1993. *The Internet for Dummies*. San Mateo, CA: IDG Books World Wide, Inc.
- Kehoe, Brendan P. 1993. *Zen and the Art of Internet: A Beginner's Guide to the Internet,* 2nd ed. Englewood, NJ: Prentice Hall.

Once users have access to the Internet, additional guides and references can be found.

JOB LISTINGS AND DIRECTORIES

America's Federal Jobs, JIST Works, Inc., 720 North Park Avenue, Indianapolis, IN 46202–3431

A comprehensive guide to more than 300,000 new job openings each year in the federal government. Gives descriptions of 150 agencies including typical jobs and requirements. Covers job grades, salary, and benefits (paperback $14.95).

Best's Insurance Reports, Property and Casualty, Best Company, Ambest Road, Oldwick, NJ 08858

Gives in-depth analyses, operating statistics, financial data, and officers of more than 1300 major stock and mutual property–casualty insurance companies. In addition, provides summary data on more than 2000 smaller mutual companies and on 300 casualty companies operating in Canada.

Chamber of Commerce Directories

Many city and area chambers of commerce publish directories that are similar to the state industrial directories but geographically restricted to areas they serve. These can normally be acquired at nominal cost.

Corporate Jobs Outlook, P.O. Drawer 100, Boerne, TX 78006, (1–800) 325–8808

Monthly publication profiling 20 companies per month. A cumulative index is sent with each month's issue.

Directory of Directories, Gale Research, Inc., 835 Penobscot Building, Detroit, MI 48226–4094

A guide to more than 10,000 business and industrial directories, professional and scientific rosters, directories, databases and other lists, and guides of all kinds. The directory is divided into 15 major classifications with more than 2100 subject headings including industry, business, education, government, science, and public affairs.

Dun & Bradstreet Million Dollar Directory—Volume I, Dun & Bradstreet, Inc., 99 Church Street, New York, NY 10007

This directory is similar to *Standard & Poor's Register*; however, it is one volume consisting of corporations with sales of $1 million or above. It is useful to use both *Standard & Poor's* and *Dun & Bradstreet* together. One directory may include a firm that the other does not, as well as additional descriptions or products, subsidies, and officer titles.

Dun & Bradstreet Reference Book of Corporate Managements, Dun & Bradstreet, Inc., 99 Church Street, New York, NY 10007

Contains data with respect to directors and selected officers of 24,000 companies with annual sales of $10 million or more or 1000 or more employees. Information given includes dates of birth, education, and business positions presently and previously held; for directors who are not officers, supplies their present principal business connection.

Encyclopedia of Associations—Volume I, National Organizations of the United States, Gale Research, Inc., 835 Penobscot Building, Detroit, MI 48226

A guide to 14,000 national and international organizations of all types, purposes, and interests. Gives names and headquarters addresses, telephone numbers, chief officials, number of members, staffs, and chapters, descriptions of membership, programs, and activities. Includes a list of special committees and departments, publications, and a three-year convention schedule. Cross-indexed.

Useful in locating placement committees that can help you learn of specific job openings in your field of interest; getting membership lists of individuals in order to develop personal contacts; learning contacts; learning where and when conferences are being held so that you can attend them and develop contacts and position leads.

Job Hunter's Source Book, Gale Research, Inc., 835 Penobscot Building, Detroit, MI 48226–4094
A new publication that lists not only profiles of professions and occupations but also information about new companies.

Job Seekers Guide to Public and Private Companies
Approximately 15,000 companies listed. Human resources and corporate officials' names given. Volumes available for the West, Midwest, Northeast, and South.

National Directory of Nonprofit Organizations, The Taft Group, 12300 Twinbrook Parkway, Suite 450, Rockville, MD 20852
Lists more than 167,000 nonprofits in the United States with reported annual income of more than $100,000.

National Trade and Professional Associations of the United States, Columbia Books, Inc., 1212 New York Avenue NW, Suite 330, Washington, DC 20005

Research Centers Directory, Gale Research, Inc., 835 Penobscot Building, Detroit, MI 48226
A guide to more than 12,000 university-related and other nonprofit research organizations established on a permanent basis and carrying on continuing research programs in agriculture, astronomy and space sciences, behavioral and social sciences, biological sciences and ecology, business and economics, computers and mathematics, education, engineering and technology, government and public affairs, humanities and religion, labor and industrial relations, law, medical sciences, physical and earth science, and regional and area studies.

Standard & Poor's Register of Corporations, Directors and Executives—Volumes I–III, Standard & Poor's, 25 Broadway, New York, NY 10004, (212) 208–8702
A guide to the business community, providing information about public companies of the United States.
Volume I—Corporate Listings. Alphabetical listing by business name of 55,000 corporations including addresses, telephone numbers, names, titles of officers and directors, public firms' Standard Industrial Classification (SIC) codes (for company industry cross-referencing), annual sales, number of employees, some division names, and principal and secondary business.
Volume II—Directors and Executives. Alphabetical list of 70,000 individuals serving as officers, directors, trustees, and partners, their princi-

pal business affiliations with official titles and business addresses. Where obtainable, year and place of birth, college, year of graduation, and fraternal memberships are listed.

Volume III—Indexes. Divided into color-coded sections:

- *Section 1: Standard Industrial Classification (SIC) Index* (green pages)
- *Section 2: Standard Industrial Classification Codes* (pink pages)
- *Section 3: Geographical Index* (yellow pages). Lists companies in the register by state and major cities. Business names are alphabetical.
- *Section 4: Cross-Reference Index* (blue pages). Alphabetical index of subsidiaries, divisions, and affiliated business units. Cross-referenced to parent company.
- *Section 5: New Individual Additions* (buff pages). Alphabetical list of individuals appearing in the Register for the first time, along with their principal business connections and business addresses.
- *Section 6: New Company Additions* (buff pages). Alphabetical list of companies appearing in the register for the first time; supplies business addresses.

Thomas' Register of American Manufacturers—Volumes 1–12, Thomas Publishing Company, 461 Eighth Avenue, New York, NY 10001

Useful in locating many specific product manufacturers, large and small, not listed in *Dun & Bradstreet* or *Standard & Poor's.*

- *Volumes 1–7:* Products and services listed alphabetically.
- *Volume 8:* Company name, address, and telephone numbers listed alphabetically with branch offices, capital ratings, and company officials.
- *Volumes 9–12:* Catalogues of companies listed alphabetically and cross-indexed in the first eight volumes.

Value Line Investment Survey, Arnold Bernhard & Company, Inc.

This service analyzes 1700 companies, domestic and foreign. Company reports (single page) with a lot of financial data. Provides good earnings and performance forecasts. Found in most libraries.

Directory of Graduate Research, American Chemical Society, 1155 16th Street, NW, Washington, DC 20036. $60 through the ACS Distribution Office, (202) 872–4405

Lists master's and Ph.D. degree-granting departments of chemistry, chemical engineering, biochemistry, medicinal and pharmaceutical chemistry, clinical chemistry, and polymer science in the United States and

Canada. Includes names of faculty members, biographical data, research interests, and titles of their recent publications.

Directory of American Research & Technology, R.R. Bowker, a division of Reed Publishing, 121 Chanlon Road, New Providence, NJ 07974

This directory includes all known nongovernment facilities currently active in any commercially applicable basic and applied research, including development of products and processes. Most of the entities are owned and operated by corporations, but some university, foundation, and cooperative organizations that do research for industry are also listed. This directory also includes a geographic index, personnel index, and a classification index to R&D activities.

Directory of Chemical Producers, SRI International, 333 Ravenswood Avenue, Menlo Park, CA, (415) 859–3627

American Consulting Engineers Council Membership Directory, 1015 15th Street, NW, Suite 802, Washington, DC 20005, (202) 347–7474, $140

ACS CAREER SERVICES

ACS Career Services refers to programs, services, and publications offered by the American Chemical Society to enhance the professional and economic development of chemists. Services, publications, and professional development programs are available at no or nominal charge to all ACS members—full members and national and student affiliates—who are seeking employment or career assistance. ACS Career Services falls under five categories, career assistance, employment services, work force analysis, publications, and workshops and presentations.

Career Assistance

- *Career Consultant Program*—Approximately 50 consultants are available to ACS members to assist them with various aspects of employment and career development. Consultants will discuss such topics as job search strategies, career transitions, interviewing techniques, employment trends, salaries, résumés, and networking. This service is available to all members at all career stages.

- *Member Assistance Program*—This program is for members who feel they've been treated in an unfair manner by their employers, colleagues, or another organization. Member Assistance cases can involve things such as denial of disability benefits, age or sex discrimination, publishing rights, termination settlement, poor performance, and personality conflict.

- *Telephone Assistance*—ACS staff are available to ACS members who call seeking information about salaries, employment trends, terms of termination, job and career transitions, simple résumé questions (otherwise, the member should send his or her résumé to ACS to be reviewed), and other career issues.

Employment Services

- *National Employment Clearing Houses (NECH)*—Job applicants at National and Regional Meetings have the opportunity to interview with employer representatives. National Meeting registration fees can be waived for unemployed ACS members who register as job applicants at NECH; registration fee waivers for unemployed members registering for NECH at Regional Meetings are at the discretion of the region. Registration for NECH itself is free at both National and Regional Meetings. There is a special résumé section for NECH for unemployed ACS members unable to attend the meetings.

- *Professional Data Bank*—A computerized registry of scientific professionals seeking employment. This service makes available information on these individuals to interested employers who seek assistance from ACS for their recruiting needs. The Data Bank is not confidential, operates on a year-round basis, and is free to ACS members and student and national affiliates.

- *Confidential Employment Listing Service*—Offers the same service as the ACS Professional Data Bank. However, for a modest charge, ACS members can have their confidentiality ensured.

- *CHEMJOBS USA*—A weekly bulletin that contains classified ads seeking the chemical professional from up to 20 newspapers and publications throughout the United States. The ads are abstracted in an easy-to-read format. A three-month subscription to this publication costs $40.

- *C&EN Situations Wanted Ads*—Employed ACS members and student affiliates may place an ad with Centcom, ACS's advertising agency, at 90 cents per word per insertion, no minimum charge. Unemployed

ACS members, student affiliates, and retired members may place free situation wanted ads; certain restrictions apply.

Work Force Analysis

ACS conducts research and produces publications on the employment and economic status of chemists. Work force publications include:

- *Salary Survey*—Analyzes the employment and salaries of ACS members who have been employed for two or more years. Condensed versions are published in a July issue of *C&EN*.

- *Starting Salary Survey*—Analyzes the employment and salaries of new chemistry and chemical engineering graduates. Condensed versions are published in an October issue of *C&EN*.

- *Workforce Report*— Published three times per year; addresses professional topics in the workplace.

- *Office of Professional Services Bulletin*—Reports present data on degrees and employment in the chemistry labor force.

- *Domestic Status, Discrimination, and Career Opportunities for Men and Women*—Presents detailed results of a 1991 survey of 4200 men and women chemists on the effects of domestic status and discrimination on the careers of men and women chemists.

- *Women Chemists*—Every five years, the ACS produces a supplemental report on the economic status of women in the ACS. Reports are available for 1975, 1980, 1985, and 1990.

- *Economic and Professional Status of Retired Chemists*—Presents results of a 1993 survey of 3500 ACS members between the ages of 55 and 70.

Publications

- *Professional Employment Guidelines*—Addresses both employer and employee good employment practices as the basis of sound professional relations. Topics include terms of employment, professional development, termination conditions, patent rights for inventors, and continuing education.

- *Academic Professional Guidelines*—These are intended to provide assistance on special issues of concern to chemical scientists in the academic environment. Guidelines deal with the obligations and responsibilities of individuals in the three segments of academic life that are important for chemical professionals: graduate students, professors, and administrators.

- *Coping with Job Loss*—This brochure describes the trauma of termination and provides information on coping with the emotional, practical, and professional aftermath. Examines the grieving process, reviews sources of help and support, and makes recommendations on organizing a job search.

- *Tips on Résumé Preparation*—Discusses different types of résumés and includes samples of each.

- *What a Ph.D. Chemist Should Consider Before Accepting a Position*— Discusses important issues a Ph.D. chemist should consider before accepting a new position: compensation, benefits, and career growth, to name a few.

- *What a B.S. Chemist Should Consider Before Accepting a Position*— Discusses important issues a B.S. chemist should consider before accepting a new position: compensation, benefits, and career growth, to name a few.

- *ACS Career, Employment & Professional Resources*—This brochure lists all ACS career resources for high school and college students exploring career options; professionals seeking employment in chemistry and allied fields; and individuals facing the challenges of career development, career changes, and retirement.

- *Younger Chemists Committee Newsletter*—Mailed twice yearly to all ACS members age 35 and younger, the newsletter regularly features articles on career development, job search, and career opportunities.

- *Teaching Chemistry to Students with Disabilities*—Developed to help professors accommodate the needs of students with disabilities in chemistry classes and labs.

- *Guide to Industrial and Academic Employment for Foreign-Born Students*—A compilation of career, employment, and resource information for foreign-born students looking for industrial or academic employment.

- *Employer Mailing List*—A mailing list used to solicit employers for ACS employment services; it is arranged by state and can be purchased for a small fee. Use of this mailing list is restricted to personal use only.

Workshops and Presentations

- *Conducting an Effective Job Search*—Discusses job search strategies, values, skills assessment, résumés, interviews, and more. Offered at National and Regional Meetings. Also offered as four separate modules: Résumé Preparation, Interviewing, Targeting the Job Market, and Job/Career Transitioning.

- *Chemistry Employment: Looking Toward the Future*—Experts from industry, academe, and government explore the political, technological, and economic trends as well as the internal organizational changes that will influence chemists' employment and the types of skills needed by chemists over the next decade. Offered at National Meetings.

- *Strategies for Successful Job Transitions*—Discusses techniques and ways to make career transitions within and outside of chemistry. Also discusses résumés, networking, values, and skills assessment. Offered at National Meetings only.

- *Chemical Technicians: Expanding Your Career Options*—Discusses work force trends, salaries, employment and education data, values, skills, transitions, and interviewing. Offered at National Meetings and by request at Regional Meetings.

- *Recruiters Panel*—A panel of recruiters from different areas of chemistry discusses what they look for in a candidate and hiring trends. Offered at National Meetings only.

- *Résumé Review and Career Assistance*—Career consultants provide one-on-one assistance to members to review their résumés and discuss various issues concerning their careers. Offered at National Meetings and some Regional Meetings.

- *Mock Interview Sessions*—Participants have the opportunity to videotape a practice interview and receive feedback from ACS personnel professionals. Offered at National Meetings only.

- *Job Security and Employment Services*—Covers employment outlook and how chemists can take control of their own careers. Offered at Local Section Meetings.

Variations of these programs are offered at local sections and on college and university campuses.

ℴℴ

APPENDIX II

TRANSITION BIOGRAPHIES

Interviews with chemists tell us that transitions may take any number of paths and may come about for a variety of reasons. Each transition is unique because of the different combination of skills, knowledge, and personal factors brought to the situation at that given time. In this section, we provide examples of how some chemists made their own successful career transitions. Each of the biographies highlights different events that brought about career transitions and different personal factors that contributed to new career paths. You, too, will discover the uniqueness of the process as you work through your own transition. By highlighting events and personal factors that contributed to the transitions of some of your fellow chemists, we hope that you will gain insight into some of the techniques that may be useful to you.

ESTHER A. H. HOPKINS

Esther Hopkins was trained as a biophysical chemist and received her doctorate from Yale University. She has made several successful career or job transitions, but the first occurred at the time she entered graduate school. When she entered college, it was with the express purpose of becoming a medical doctor. She completed her undergraduate work as a premedical student and promptly applied for entry into medical school.

Dr. Hopkins was not selected for medical training. For some, the reaction might have been to forgo any further education. Dr. Hopkins,

3052–8/95/0163$08.00/0 ©1995 American Chemical Society

however, chose to begin graduate training in chemistry, a subject that she confesses was always one of her favorites. Dr. Hopkins completed her Ph.D. at Yale and was offered several positions after graduation, finally selecting a position at Polaroid Corporation as a physical chemist.

Dr. Hopkins remained at Polaroid until she chose early retirement, but not before she had made yet another significant career transition. An appointment to a Polaroid committee dealing with South African concerns brought her once again to another turn in the road. While serving on the committee, she became acquainted with another committee member who was studying law, and after further conversations, Dr. Hopkins began to consider a career in patent law.

Dr. Hopkins weighed all of the factors associated with beginning a new career and discussed the study program with her family. Her next step was to apply for tuition assistance from Polaroid, which she received. A strong personal need for accomplishment enabled her to surmount difficult obstacles, including convalescence after an automobile accident. Following her graduation from law school, Dr. Hopkins transferred into the New Business Development area at Polaroid, and remained there until her early retirement.

This was not the last of Dr. Hopkins's transitions. With the help of her active network, she took a position with the Massachusetts State Environmental Office reviewing contracts. She does not use her science background every day, although she does on occasion help others by providing assistance on scientific matters. This contract law position has given Dr. Hopkins an opportunity for a new learning experience and broadens her work options for the future.

DAVID ELROD

Ever since the ninth grade, David Elrod knew that he wanted to be a chemist. In his undergraduate program at Kalamazoo College, he decided to major in organic chemistry. After graduation, he joined the Upjohn Company as a chemistry associate in the Cancer Research Department performing discovery, isolation, synthesis, chemical modification, and biosynthesis of natural products. During his 12 years' work in this field, he progressed up the chemistry associate ladder, coauthoring seven papers and seven patents.

Dr. Elrod was encouraged to further his chemistry education and completed his master's degree in organic chemistry at Western Michigan University.

In late 1985 several events conspired to make Dr. Elrod consider a career shift from laboratory chemistry to computer chemistry. An internal reorganization, he believed, was likely to lead to the cessation of natural products screening for cancer drug delivery; his supervisor was rumored to be considering retirement; and an internal job posting from the newly formed Upjohn Computational Chemistry Department appeared, which advertised for someone with a strong knowledge of organic chemistry who was able to do occasional computer programming.

Dr. Elrod had taken two computer programming classes at Western Michigan as part of the research tool requirements for a Ph.D. degree in chemistry. As a result, he had the qualifications for the computational chemistry position, and he was hired.

When Upjohn later needed someone to take responsibility for the DNA and protein-sequence analysis software in his new department, Dr. Elrod volunteered. This post led to an opportunity to write a menu-driven interface for the multitude of sequence analysis programs. Upon the enthusiastic recommendation of his colleagues, Dr. Elrod attended a short course on neural networks and returned excited about new computational technology. During this time Dr. Elrod met a computer science professor from Western Michigan who was spending one semester of his sabbatical at Upjohn. This meeting led Dr. Elrod to collaborate on several papers on applying neural networks to the prediction of chemical reactions and helped him to earn a Ph.D. in computational chemistry at Western Michigan.

Dr. Elrod believes that his successful new career direction, merging chemistry and computers, developed from an old marketing adage which states that to be successful, you must "find a need and fill it." He thinks this is true with many careers, chemistry included. Given our technologically complex society, there is a real need for people with a strong understanding of chemistry in many careers that do not directly involve laboratory work.

ELINOR OWENS

Elinor Hankins Owens received her Ph.D. in organic chemistry from the University of Wisconsin. Following graduation Dr. Owens joined The Rohm and Haas Company, where she worked in a laboratory devoted to the synthesis of new monomers. She soon became an expert in the field of adhesion-promoting monomers and obtained a number of patents in the field. After 10 years in this area, Dr. Owens transferred to the library, where she was appointed editor of *Developments in Chemistry,* a current awareness bulletin.

Shortly after this transition to the library, Dr. Owens married. This was an era in which many women did not combine work and raising a family, and Dr. Owens decided to "retire." It was her decision to remain at home, and as a result, she did little to maintain her knowledge of chemistry or chemical information. In 1978, Dr. Owens decided to return to the job market. She realized that because she had been out of the laboratory and the library for 15 years, she lacked the up-to-date skills and knowledge for full-time employment. Because her children were still in school (her youngest was nine years old), she decided that part-time employment would be the best alternative. Dr. Owens took a part-time job with a local analytical laboratory that specialized in the analysis of food. The job suited her well as she had always enjoyed laboratory work.

The situation that Dr. Owens faced is one still faced by many today: how to balance a family and a career. For the chemist who remains at home and cannot or does not keep technical knowledge up-to-date, a return to the job market on a part-time basis is one alternative.

PAUL Y. FENG

Paul Feng has made a number of major career moves—industry, contract research, academe, law, and business. He has a Ph.D. in chemistry from Washington University, a law degree from De Paul University, and an MBA from the University of Chicago. In 1988, he negotiated early retirement from a tenured full professorship at Marquette University, but still remains close to education by teaching chemistry part-time at National-Louis University in Evanston, Ill. He is "Of Counsel" to two law firms in addition to working for his own law office. With his colleagues, he works on intellectual property, business litigation, immigration, and general practice. He has been admitted to practice at the Patent and Trademark Office, the Patent Bar, several federal and state courts, and is active both in his local bar association and the American Chemical Society.

Dr. Feng believes that his career moves can best be characterized as diversification rather than total changes. He also believes that the moves were rather smooth and credits the smoothness to luck, a belief in the value of his endeavors, preparation, and above all, help from numerous teachers and colleagues at various universities, government, and industrial laboratories.

In the move from industry and contract research to academe, Dr. Feng benefited from his experience in part-time high school and college-level teaching. He used his accumulated vacation time to obtain this relevant professional experience.

Dr. Feng's interest in the legal profession has a long history. While growing up in China, Dr. Feng frequented the Beijing Court, which was then headed by one of his close relatives. He received his law degree from De Paul's evening program in 1986 and passed the Illinois bar exam the same year. While waiting for the results of the bar exam, Dr. Feng spent the summer of 1986 at the Criminal Court of Cook County (Illinois), where he first assisted and then conducted criminal trials under a special program in the courtroom of Judge Earl Strayhorn—an effort which gained him early admission to the Federal Trial Bar. Subsequently, he served as a volunteer attorney on immigration matters for the Travelers and Immigrants Aid in Chicago. He also performs *pro bono* legal work through a Chinese church.

In business matters, Dr. Feng carried out research for many years on corporate dividend policy and other corporate financial data for his personal use. He also sold his research compilation on dividend reinvestment to major corporations and other institutions. His formal move in "career diversification" in this area was enrollment in the Graduate School of Business at the University of Chicago in 1989. Dr. Feng was one of 12 MBAs with Honors out of a class of 555 in the class of 1991. He found the training from his program to be helpful in his legal work (business transactions and litigations), in science (better knowledge of statistical methods), in financial planning, and in his duties as a member of the Board of Managers of the North Suburban Bar Association in Chicago.

In view of the current state of the economy, Dr. Feng believes that a professional cannot ignore the possibility of a voluntary or involuntary change in career. Dr. Feng believes that a person should evaluate honestly the pluses and minuses of any contemplated new endeavor. If the evaluation leads to a "go" decision, the person has to believe in himself or herself, learn from the past and the present, and plan for the future.

DORIT L. NOETHER

Dorit Noether has pursued several careers, each of which has been enriching. In 1938, having completed only the 10th grade in her native Austria, she had to leave high school due to the war. More than two years later she landed with her family in Boston and entered Radcliffe College as an upper sophomore. Louis Fieser's course in organic chemistry determined her future: She would become an organic chemist. She was married before beginning her senior year and after graduation in 1943 moved to New York, where her husband Herman joined Celanese Corporation as a research chemist.

Dr. Noether started graduate school at Columbia University but decided to interrupt her studies in 1947 shortly before the birth of her second child. For the following 10 years she was a full-time mother to her three children. To keep her chemistry up-to-date she served as a freelance abstractor for Chemical Abstracts Service. When her youngest child entered first grade, she obtained a part-time teaching assistantship at Adelphi College, and the following year, she obtained an instructorship at the newly formed C.W. Post College. She was the only chemist at Post and had to run lectures, labs, and staff the stockroom. After two years she requested a promotion but was denied it because she didn't have a Ph.D.

By now she was 37 years old. She entered graduate school at Rutgers in the fall of 1959. In 1960 there was a faculty opening and she obtained an instructorship, which was difficult for a woman in those days. Prior to that time, women had not even held a teaching assistantship in the department. Dr. Noether earned her doctorate in 1964, was promoted to assistant professor, and stayed at Rutgers for 10 years.

She then stopped teaching and became a special consultant for Master Planning at the New Jersey Department of Higher Education in Trenton. Her years at the department taught her about university administration and gave her extensive opportunity for writing. When CHEMTECH magazine needed an associate editor, Dr. Noether embarked on a new career that lasted for 16 years.

In 1993, Dr. Noether left CHEMTECH to transform an avocation into a career: She hopes to become a better watercolorist.

APPENDIX III

CAREER AREAS FOR CHEMISTS

Industry	Products or Areas
Agencies and contracts	State and local waste and forensics; federal agencies (such as NSF, NIH, Navy laboratories, Air Force laboratories, ORNL)
Agriculture and fertilizer	Fertilizers, pesticides, herbicides, wetting agents, garden chemistry
Audit	EPA, OSHA, FDA, Inspector General, state regulations, GLPs, MSDS
Biomedical products	Ultrapure materials, prostheses, mono- and multicellular processes
Clinical	Radiochemistry, selected pharmaceuticals, hospitals, service labs, government labs, protocols, methods
Consultants	See listings in the *Directory of Scientific and Technical Consultants and Expert Witnesses* (American Society for Testing and Materials: Philadelphia, PA, 1994), *C&EN*, *American Bar Association Journal*, university staff, ACS Experts Roster (Government Relations and Science Policy Office 800–227–5558)

3052–8/95/0169$08.00/0 ©1995 American Chemical Society

Industry	*Products or Areas*
Consumer products	Soaps, detergents, cleaners
Consumer "sin" products	Alcohol, wine, beer, tobacco
Energy	Petroleum, coal, solvents, oils and greases, lubricants, fuels, solar materials, battery material, exploration and recovery, fuel cells
Food products	Fragrances, spices, foods, flavors
Forensics	Specialized techniques (microscopy, sampling, mass spectroscopy, chromatography, residue analysis, microbiology technology)
Industrial products (inorganic)	Peroxides, acids, bleaches, pigments, colloids, ceramics, cements, catalysts
Industrial products (organic)	Solvents, polymers, paints, plasticizers, extenders, resins, rubber, dyes, cellulose and paper, colloids, garden chemicals, specific fuels, textiles, catalysts
Instrumentation development	All analytical techniques, including physical science, medical science, real-time, on-line, process control
Law	Lawyers, paralegal, patent examiners, patent agents, chemical information
Marine science and geochemistry	Fish and plant chemistry; contaminations, historical evidence, oceans, rivers, subsurface, volcanoes
Mining and metallurgy	Corrosion, leaches, physical properties, waste, upgrades
Nuclear and radiation	Government laboratories, government contract laboratories (such as ORNL, Los Alamos), Superfund areas, selected universities (e.g., University of Kentucky)
Personal care	Toothpaste, body products, deodorants

Industry	Products or Areas
Pharmaceuticals and medical products	Over-the-counter drugs, prescription drugs, molecular biology, computer modeling, analoging, radiochemistry, drug discovery, natural products, animal drugs, animal health
Polymer products	Diapers, towels, clothes, wall coverings
Quality and mathematics	GLP (good laboratory practice), sampling, applications, ISO 9000, chemometrics
Rubber	Tires, belts and hoses, apparel, coverings
Standards and methodology	All analytical techniques, including government regulations, physical science, medical science
Ultrapure materials	Semiconductors, optical fibers, diamonds, phosphorus
Universities and chemistry departments	Research, teaching, service labs, chemical information, local outreach
Waste management	Air, water, soil, surface, subsurface, dumps, spills, federal regulations, state regulations, local system analysis
Weapons and defense	All analytical techniques, including physical science, medical science, chemical warfare, treaty verification

INDEX

Production: Paula M. Bérard
Acquisition: Cheryl Shanks
Book design and typesetting: Betsy Kulamer
Cover design: Dick Hannus

Printed and bound by Maple Press, York, PA